# INSPECTIONS IN HAZARDOUS AREAS

## Ian Staff

**ELECTRICAL TRAINING CONSULTANT**

First Edition published 2020

2QT Limited (Publishing)

Settle

North Yorkshire

BD24 9RH

Cover design: Dale Rennard

Images supplied by author

Additional images from iStockhoto.com

Printed in the UK by Lightning Source UK Limited

ISBN 978-1-913071-61-5

# Introduction/Preface

I am an Electrical Training Consultant for a training company in Hull by the name of Humberside Offshore Training Association Ltd (HOTA), Malmo Road. Before my fifteen or so years at HOTA as a trainer/assessor I spent thirty-eight years with BP, seven of those years as their Instrument/Electrical Training Officer in charge of all instrument and electrical training in their training complex where I obtained my training and assessing qualifications. HOTA have the franchise to several courses under the heading of 'CompEx' which is a globally recognised, competency based, set of training packages with JTL Ltd as the national body in overall control of the CompEx Scheme. I carry out some training and assessments in a range of CompEx courses.

Following on from my last book 'Hazardous Areas for Technicians' (ISBN: 978-1-91204-95-8) I have produced this book 'Inspections in Hazardous Areas' both of which complement the essential EEMUA manual (ISBN: 978-0-85931-212-7) This book specialises on Inspections by giving spreadsheets of Visual, Close and Detailed points to be checked on protection systems: Exd, Exe, Ext, Exn, Exi, Exp and Mechanical. The applicable International fault codes are included against the faults on the spreadsheets, and in the back of the book is **my** interpretation as to what the fault codes refer to and why that particular check is being carried out. So if we take a lighting circuit, the Inspector could look up the spreadsheet: Visual, Close or Detailed, and it would give them all the faults that they were looking for and the International fault codes that went with them. They could then look up the code further back in the book for explanations and diagrams showing the importance of why that particular check was carried out.

> Ian Staff SB St J.
> Electrical Training Consultant.

## Please Note

This book should only be treated as an <u>advisory guide</u> for the inspection of equipment in a hazardous area and used in conjunction with the EEMUA Manual and current legislation, Standards and Codes of Practice. It does not suggest overriding company policies and procedures in any way. There is a list of advisory publications below:

## Recommended Publications

1 – The EEMUA Practitioners Handbook (ISBN: 978-0-85931-212-7)

2 – IEC60079 Explosive Atmosphere Standard – Section 17

3 – ISO80079 Explosive Atmospheres (Non Electrical Equipment Exh)

4 – Hazardous Areas for Technicians (ISBN: 978-1-912014-95-8)

# Humberside Offshore Training Association (HOTA)

HOTA was established in 1987 and based in Hull, East Yorkshire and is a Quality Training Provider offering nationally approved training and tailor-made bespoke courses to meet individual and company specific training needs.

A Limited Company with Charity Status, all surplus funds generated are invested back into enhancing training and delegate facilities at The Cullen and The Ellis Buildings – Malmo Road and its Albert Dock Site.

HOTA is renowned for its flexibility, professionalism and its industry experienced, highly trained team of trainers delivering courses when, where and how they are required.

Since 1999, HOTA has been offering the JTL Approved Electrical Equipment in Hazardous Areas (CompEx) Course and over the years has continually added to the existing Training Portfolio. HOTA now offers the CompEx: Main Course (EX01 – EX04) CompEx Refresher (EX01R – EX04R), CompEx Foundation (EXF), CompEx Dust (EX05 – EX06) and CompEx Mechanical Equipment in Hazardous Areas (EX11).

HOTA's assessment suite is a world class facility and delegates are allotted their own fully equipped assessment bay. All courses are conducted by highly experienced JTL approved trainers and assessors.

To complement the CompEx training courses, HOTA conducts various City & Guilds Approved Electrical Courses as well as bespoke training courses including the one day Introduction to Hazardous Areas and pre CompEx Glanding Practice.

HOTA provides Nationally Approved and Bespoke Electrical and Mechanical Training Courses including:

## JTL Approved:

- CompEx EX01 – EX04 Electrical Equipment in Hazardous Areas and Explosive Atmospheres (five days)
- CompEx EX01R – EX04R Refresher after 5 Years (three days)
- CompEx EX05 – EX06 Electrical Equipment in Dust Atmospheres (three days)
- CompEx EX11 – Mechanical Equipment in Hazardous Areas & Atex. (three days)
- CompEx EXF – Foundation Course (two days)

## City & Guilds Approved:

- IEE 18th Edition 2382-2-18 Level 3 Wiring Regulations.
- PATS 2377-22 Level 3 Award In-Service Inspection & Testing of Electrical Equipment.
- 2391-02 Level 3 Inspection & Testing.
- City & Guilds 2919-01 Level 3 Award in Domestic, Commercial and Industrial Electric Vehicle Charging Equipment Installation.

## Bespoke Training:

- Introduction to Hazardous Areas
- Pre-CompEx Glanding Course
- Mechanical Joint Integrity
- Atex Familiarisation
- Electricity at Work Regulations – Practitioners and Non-Practitioners
- Static Electricity in Hazardous Areas.

Offshore Oil & Gas Training Courses

Maritime Training Courses

Wind Turbine and Renewable Energy Training Courses

Emergency Rescue Response Vessel Training Courses

CompEx Electrical and Mechanical Training Courses

City & Guilds Training Courses

Medical – First Aid Training Courses

Health & Safety Training Courses

Firefighting Training Courses

Specialist Training Courses

Royal Yachting Association Training Courses

For a full list of courses and dates please visit HOTA's facilities:

Website: www.hota.org

Telephone: +44(0) 1482 820567

Email: bookings@hota.org

Address: Malmo Road, Sutton Fields Industrial Estate, Hull. HU7 0YF

HOTA's purpose-built training facilities have ample free onsite parking and free Wi-Fi throughout the site.

Delegates can also enjoy a free two course lunch and refreshments in the onsite 100 seat restaurant.

# Contents

## Documentation:

## Visual Inspection:

## Close Inspection:

# Detailed Inspection:

# Power Code Descriptions:

# IS Code Descriptions:

# Pressure Code Descriptions:

# Mechanical Descriptions:

# How to use the Book for Inspections:

1   Read the section at the front describing what comprises an Inspection.

2   Read the 'Competency of Inspectors' Section.

3   Look at the documentation required by the Company Policy to complete the Inspection. (Permit, Gas Free Certificate etc. and decide which is applicable.)

4   Decide on the 'Protection' of the equipment e.g. Exd, Exe, Exn, Exi, Exp or mechanical.

5   Turn to the Visual, Close or Detailed **Checklist** associated with that protection.

6   Look down the points and codes that apply to the equipment being inspected.

7   If you want to know the meaning of any code, look at the code descriptions further back in the book. These are **MY** definition of each of the codes under the different protections of Exd, Exe, Exn, Exi, Exp or Mechanical.

# What is an Inspection?

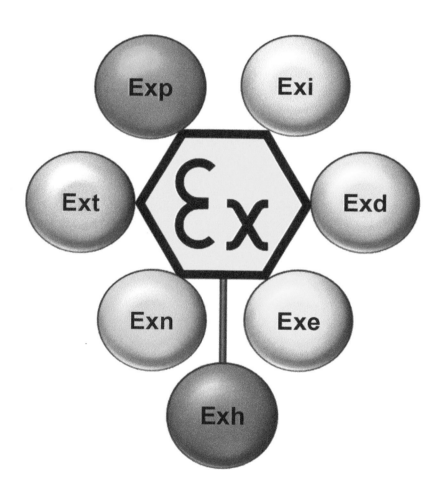

# Hazardous Area Equipment

One way to look at inspecting electrical, instrument and mechanical equipment in hazardous areas is as follows:

Think of the equipment as a 'box with a lid' as in the diagram on the right. Inside the box is the electrical/electronic/mechanical contents (shown in blue).

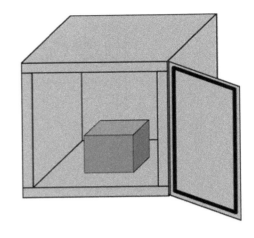

Sometimes these contents are sparking (e.g. Exd & Exp is required), sometimes they are non-sparking, (Exe, Exn, Ext & Exh will suffice) and sometimes the spark is not strong enough to ignite a gas (Exi).

The box has a temperature which it must not go over under normal and specified fault conditions: 'Temperature Class'.

The box has a seal on the lid which will determine both the water ingress protection (IP) and whether gas, vapour or dust can impregnate (Exe, Exn, Exi, Ext & Exh).

Exd flameproof equipment will allow gas to enter without compromising safety because the box is strong enough to contain an explosion in the gas group range to which it is designed. The box has flameproof paths to cool explosion pressure so that the hot explosion gases cannot propagate outside the box.

Sometimes we pump an inert gas (air) into the box to create a positive pressure thus keeping any gas, vapour or dust out (Exp). We can quench the arc by filling the box with a medium of oil or powder (Exo, Exq or Exh). Sometimes the box may be mechanical so may be Exh, Exd or Exp.

The box is certified to a certain Standard by a 'Notified Body' and a whole range of hexagonal shapes, letters and numbers are put onto a Hazardous Data Plate to ensure that the box complies, fully, with the information above and will do exactly what the manufacturer claims, (given 'Protection' & 'Gas Group' & 'Temperature Class' etc).

To work on this box Technicians and Inspectors have to be able to prove competence by understanding what the above information translates to and what these hexagonal shapes, letters and numbers stand for and be able to determine if the box is safe. (CompEx Competency Course.)

Inspecting the box at specified intervals, usually three years, will determine if it remains safe through the years and does not succumb to any damage, modification or wear that may make it unsafe.

This book sets out to firstly let you know what to inspect and then why we do that particular Inspection, and centres on six forms of electrical 'Protection' namely Exd flameproof, Exe increased safety, Exn reduced risk, Ext protection by enclosure (Dusts), Exi intrinsic safety, Exp pressurisation and Exh mechanical Inspections.

# Inspector Competence?

# Inspector Competence

Sometimes the electrical, instrument or mechanical technicians carry out Inspections as routine maintenance on their plant. They may not just do Inspections as they may have a whole range of other tasks. Other companies may enrol 'Inspectors' who may or may not have been technicians in the past.

Whichever of the above examples the company chooses, the people carrying out these Inspections have the **'Ability'** and must be **'Competent'** to do the task. So what qualities would a company look for?

1   Competency: How can Inspectors show competency? Well, one of the ways of proving competence is for these Inspectors to take a JTL CompEx course, passing an assessment at the end and gaining a **Competency Certificate** in EX02 (Power Inspections) & EX04 (Instrument Inspections) Gases & Vapours and/or EX06 in the Dust world. Mechanical would be the EX11.

   Centres run five-day courses where Inspections are a significant element, or many centres will run, say, three-day courses just for Inspectors with no installation module. On this JTL course Atex Markings, Zones, Gas Groups, Temperature Classifications etc. are covered.

2   Knowledge: A great deal of knowledge would be gained from courses and is vital when checking the information on the equipment. The more Inspections that are completed, the more knowledge and experience is gained.

3   Detail: Attention to detail is vital in fault finding, rather than quickly scanning over to get the job completed and missing several important faults that may put both plant and people at risk.

4   Expertise: Is gained when carrying out the work. The more Inspections that are completed the better the Inspectors become.

5   Quality: A company will become aware of the quality of the Inspectors by the work they do and the reports that they put in. A company will get to know which Inspectors they can rely on to complete the work correctly.

6   Skill: Is gained, as is expertise, by carrying out the work and knowing what sort of thing is being looked for. This all contributes to being capable of doing the work.

The scope and specification of the role of Inspector must be made abundantly clear at the beginning:

   i)      How far do they go?

   ii)     Do they carry out isolations?

   iii)    Do they carry out any repairs?

   iv)     Who do they report to?

These Inspectors must also, at intervals, be interviewed and assessed themselves to ensure that they remain updated on changes to legislation.

# Grades of Inspection

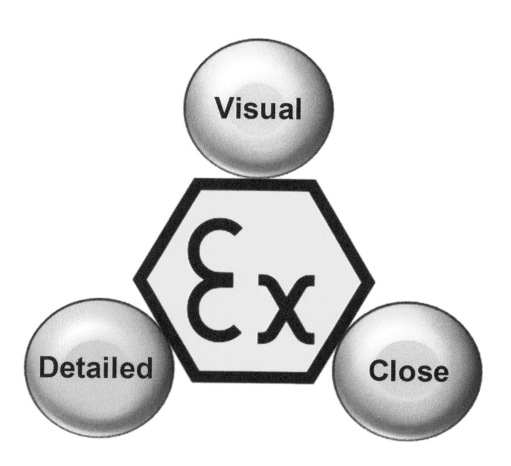

# Grades of Inspection

In the electrical & instrument world there are three grades of Inspection namely: **Visual, Close and Detailed.** In the mechanical world there are only two grades of Inspection namely: **Visual and Detailed**, missing out Close.

## Visual Inspections:

Visual inspections identify, without the use of access equipment such as steps, ladders etc., or tools, those defects such as missing bolts, stoppers etc. that will be apparent to the eye.

Binoculars for lighting circuits and thermal image cameras for, say, motor control centre unit temperature, and temperature guns would not be classed as tools.

No Gas Free Certificate required as no opening up.

These Visual Inspections can be done live. All that the Inspector/Technician has is an 'Area Classification Drawing' (a drawing with lines on showing the Zones).

## Close Inspections:

Close Inspections identify those defects such as loose glands, lids, bolts, screws, stoppers etc. Now if the Inspector/Technician has come along just to do a Close Inspection they will have to encompass those aspects covered by the Visual Inspection.

Access equipment can be used here and tools just to check tightness, not to open up, as no isolation is required. This time, as well as the Area Classification Drawing, a Loop Diagram for instruments and a Schematic for lighting is also available.

No Gas Free Certificate required as no opening up.

## Detailed Inspections:

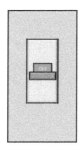

Detailed Inspections identify defects such as loose terminations, broken terminal block, too much copper showing etc.

Now if the Inspector/Technician has just come along to do a Detailed Inspection then they will have to encompass those aspects covered by the Close and Visual Inspection. Access equipment can be used here and a full **ISOLATION** procedure carried out as equipment covers will be removed.

This time, as well as the Area Classification Drawing, a Loop Diagram for instruments and a Schematic for lighting is also available.

A Gas Free Certificate is required as covers are to be opened and supply checked. Enclosures are considered to be live until proven dead, hence the Gas Free Certificate.

# Types of Inspection

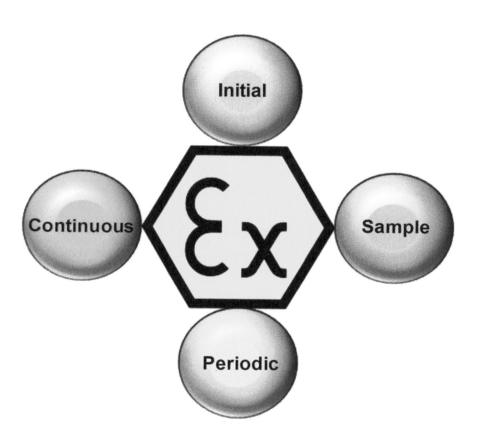

# Types of Inspection

We have discussed 'Grades' of Inspection in a previous section, now we will discuss 'Types' of Inspection. There are four 'Types' of Inspection namely: Initial, Periodic, Sample and Continuous. Let us discuss each of these types of Inspection and in what circumstances each would be used.

## Initial Inspections:

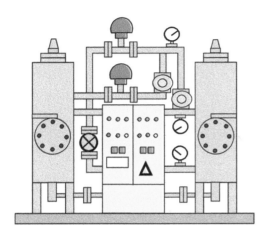

Initial Inspections are **Detailed** Inspections, some companies may call these **'Commissioning'** Inspections and these are carried out on all equipment before it is brought into service such as the drier system shown left.

Every item of equipment in a particular system is opened up and checked both internally and externally.

So what is inspected on the equipment? As well as the above checks, one of the most important things on an **Initial** Inspection is to check the 'Protection' i.e. Exd, Exe, Exn etc., to ensure that it is appropriate for the Zone as well as the gas group, temperature class etc. into which the equipment is installed. The results of these Inspections must be formally recorded and form a data baseline for future Inspections.

## Sample Inspections:

Sample Inspections, as the name suggests, are carried out to a **sample** of the equipment. If we take the plant lighting system there could be 3,500 lights. To inspect every one could take months.

On the **'Initial'** Inspection 100% of the equipment was tested, now a system could be set up, for example **'Sample'** Inspection of 10% checks. Each year the company would inspect 10% of the lighting system on routine maintenance. In ten years they will have covered 100%. In the meantime, technicians are changing lamps, process report any broken fittings that they see etc.

If there are many faults showing up on the **Sample** Inspections then a larger percentage of equipment may be inspected.

The Sample Inspection should be used mainly to monitor the effects of environmental conditions, the effects of vibration, design weakness etc. These Inspections are not carried out in order to find loose connections etc. although they might highlight such problems.

These Sample Inspections can be **Visual, Close or Detailed.**

# Periodic Inspections:

Periodic Inspections are carried out usually as part of 'Routine Maintenance' or 'Condition Monitoring' and these would, of course, be on existing fixed plant. So what is inspected on the equipment? This Inspection would identify any deterioration from the last Inspection, e.g. has anything changed?

The frequency of these inspections is very important. Too far apart and breakdowns could occur, too close together and money and resources are wasted as nothing will have changed from the last time. The '**Periodic**' Inspections should not be more than three years apart.

These '**Periodic**' Inspections must be organised and planned in with the Operations team in charge of the plant. As a Maintenance Technician, I am sure that you have, in your career, many times gone to inspect something and find that it is running and when you have gone to the Process team they are, for some reason, either unwilling or unable to shut the item down.

Some important points that may help the engineer to determine the frequency are:

1   The environment that the equipment is in. If the environment is hostile you may want to carry out inspections more frequently.

2   What was found on previous inspections? If there is a trend of the same equipment causing a problem regularly the design might require change.

3   The amount of use the equipment gets. This would have a bearing on frequency of inspections as items of equipment may wear more quickly if frequently used.

4   Routine Maintenance. As mentioned earlier, routine maintenance may determine how often the equipment is inspected. This interval time will have to be reviewed by the engineer.

5   Manufacturer's guidance. The manufacturers may stipulate how often their product should be inspected. Going under their time is no problem, but they should be consulted if you go over it.

6   Similar equipment in other locations. Other plants may be having problems with equipment similar to this. You may want to discuss their inspection frequency with them.

7   How important to the plant is the equipment? So here you would be looking at consequences of a shutdown. You might decide that very important equipment is inspected more frequently.

8   Have there been many breakdowns? If there have been a few breakdowns you might decrease the time interval between inspections or use the information gained from previous inspections to change the design.

# Continuous Inspections:

Continuous Inspections are where very skilled technicians who are very familiar with this particular equipment carry out **'frequent'** inspection and maintenance.

These technicians would be skilled in this particular electrical installation and would ensure that the equipment and its environment maintain the explosion protection features. These Continuous Inspections would be **Visual or Close.** Much of this type of inspection can be covered with condition monitoring.

Inspections may include:

1    Temperature: If we take an electric motor there is a certain temperature that the windings are geared to. The designation letter in the table below is taken from the motor data plate. There are 4 temperatures and they go up in 25C sections.

| Letter Designation: | Maximum Insulation Temperature: $^{\circ}$C |
|:---:|:---:|
| A | 105 |
| B | 130 |
| C | 155 |
| D | 180 |

2    Vibration: If we take electric motors, vibration can be caused by a number of faults that may not be easily found on an Inspection. Misalignment may be one common cause, others can be catastrophic burning out causing electrical imbalance, the load faults transmitting vibration to the motor or simply the foot bolts becoming loose, just to name a few.

3    Bearing Condition: Bearings can be checked with the motor running. There are many hand-held bearing monitors on the market.

4    Noise Level: All of the problems listed under 'Vibration' may cause the noise level to rise on the machine. If there is a rise in both vibration and noise something may be going catastrophically wrong.

5    Electrical Current: Again if we take an electric motor there are several different currents:

   a    **Start Current:** Motor can take 6 x full load current returning to run current.

   b    **No Load Current:** Motor is run separate from the load.

   c    **Run Current:** Motor is connected to the load. If this goes up, investigate.

   d    **Full Load Current:** Motor running on ammeter red line. No problem if the motor has always run there.

   e    **Overload Current:** Needle is over the red full load current line. Eventual trip.

   f    **Fault Current:** Trip or burnout.

   g    **Stall Current:** Motor back onto 6 x full load current. Quick trip/fuse blow.

# Documentation

## Documents

Permit

Clearance Certificate

Gas Free Certificate

Method Statement

Risk Assessment

Area Classification Drawing

Equipment Location Diagram

Checklists

Report on Inspection

Schematic Diagram

Loop Diagram

Piping & Instrument Diagram

# Inspections Documentation:
## (Check Company Policy & Procedures)

In all of the literature on **Visual** Inspections it states that the only document allowed/required is an **'Area Classification Drawing'.** On other Inspections Schematic/Loop Diagrams are required. Below are several extra documents that may be required on various Inspections depending upon different company procedures and policies.

1   Permit to Work: This depends solely on company policy, but the chances are that the Inspector(s) would not be allowed on the plant without one.

2   A Clearance Certificate: Some companies do not use or rely on these Clearance Certificates and just use a Permit to Work.

3   A Gas Free Certificate: If test equipment is being used, covers opened, or where there is a potentially live supply. Required on an Electrical Detailed Inspection.

4   A Method Statement: Written by the Electrical, Instrument or Mechanical Engineer stating the safe system of work which will also include the scope of work. (i.e. How far do the Inspectors go?)

5   A Risk Assessment: The Permit will go a long way towards this, but that is a company document. This Assessment is completed by the Inspector(s).

6   An Area Classification Drawing: The Inspector(s) would need to know what Zone they are in so they can ensure that the equipment is suitable.

7   An Equipment Location Diagram: The Inspectors(s) would need to know exactly where on the plant the equipment is that they have to inspect.

8   The Checklist: This would show what faults they would be looking for on an Inspection and what codes would go with those faults.

9   The Report: This is the form where the Inspector(s) write down the location, fault, code and suggested remedial action.

10   A Circuit Wiring/Schematic Diagram: If the equipment is a power circuit such as lighting, motors, sockets etc. Usually a Schematic Diagram is used on an Inspection, a Wiring Diagram would not be a great deal of use here and would be more use on an Installation Project.

11   A Loop Diagram: If the equipment is instrumentation, the Loop Diagram would be invaluable as all the information on the instruments and barrier units would be on here. This drawing should be as near 100% accurate as possible all of the time.

12   A Piping and Instrumentation Diagram (P & ID): This would be required on the Mechanical Inspections to show the pipework and instrumentation included in the inspection.

# Permit to Work

A Permit to Work is a document that allows the person appointed by the company, e.g. the 'Appointed Person' for that particular area to keep control of all work that goes on in it. Many Permits these days are 'E' Permits and completed on a computer, but hard copies are the only way that people can sign on them.

Some people look upon a Permit as a licence to work, but it is much more than that. Firstly it is a legal document that can be used in a court of law to prosecute or defend you. A Permit, if you like, can be called a legal Risk Assessment stating exactly what work is to be done and where, what isolations have been carried out - both process and electrical - to protect you, and what hazards remain that perhaps cannot be removed, and the precautions to be taken to counteract those hazards.

Once the Permit is written the person who accepts it, usually the team leader of a team or an individual if the Permit is just for one person, must accept the Permit in the presence of the authorised person who has written it, as at that point the acceptor has got what is called **'The right of challenge'** so if at that point there was something that the acceptor was not happy with it can be challenged and may be changed. Once the Permit is accepted then no changes may be made to the content.

There are usually a set number of spaces for others to sign on who are working on that particular job. If these spaces are filled, then an addendum may be stapled to the Permit with more 'signing on' spaces. I say stapled and not paper-clipped because it is much harder for it to become detached and if the Permit had a torn corner then it may be suspected that something had been attached.

Whoever accepts the Permit should be the one who signs in the section **'work is completed'** when the job is done and they have walked round and checked everything. If the job carries on for several shifts continuously then there should be a provision on the reverse for the team leader of each shift to accept the Permit.

When it is written the Permit is valid up to a certain time, after this time the permit time has to be renewed by an authorised person and again, if the job was to go on continuously for several shifts then this person will change. The acceptor must check that the Permit has been updated at the end of each time period.

Electrical de-isolations will **not** be done until the **'Work Completed'** section has been signed by the team leader of the team carrying out the work and the **'authorisation to de-isolate'** signature has been obtained from the authorised person.

Finally the Permit is cancelled by the authorised person and kept in archive for a period of time.

Sometimes on the reverse of the Permit is what is called **'sanction to test'**. This is where electrical isolations can be de-isolated temporarily and then after the test the isolation is reinstated. This can only be done once everyone named on the Permit knows that this is happening. Let us say a motor returns from rewind and the technician wants to run it first before connecting to the pump, the fuses cannot be replaced normally because a lot of work has to be done before the Permit can be cancelled.

| Permit Number:<br>1234256B | # Permit to Work | HOTA |
|---|---|---|

| 1 - Location of work being carried out: | 2 - Permit valid From time of Certification Until: | |
|---|---|---|
| | Time: | hrs |
| | Date: | |

**3 - Description of work to be carried out:**

| 4a - Process Isolations: | Time: | Date: | Initial: |
|---|---|---|---|
| 1 - | | | |
| 2 - | | | |
| 3 - | | | |

**4b - Electrical Isolations:   (Electrical Technician)**

Please Isolate the items below:          Signed:          (Authorised Person)

| Distribution Board Number: | Circuit Number: | Fuse Size: | Date of Isolation: | Time of Isolation: | Signed by Technician: | Company: |
|---|---|---|---|---|---|---|
| 1 - | | | | | | |
| 2 - | | | | | | |
| 3 - | | | | | | |

| 5 - Hazards Remaining: | 6 - Precautions to be taken: |
|---|---|
| 1 | 1 |
| 2 | 2 |
| 3 | 3 |

| 7 - Certification:  (Authorised Person) | 8 - Acceptance: |
|---|---|
| I Certify that the Isolations in Sections 4a & 4b have been carried out and the work may start subject to requirements of sections 5 & 6: | I have read and understand this Permit to Work and accept all the Hazards Remaining and the precautions to be taken: |
| Signed:                              Print: | Signed:                              Print: |
| (Authorised Person)               Time: | Position:                            Time: |
| Company:                            Date: | Company:                            Date: |

| 9 - Sign onto the permit (1 - 3): | 9 - Sign onto the permit (4 - 6): |
|---|---|
| 1                         Date: | 4                         Date: |
| 2                         Date: | 5                         Date: |
| 3                         Date: | 6                         Date: |

| 10 - Work Completed:  (Acceptor) | 11 - Electrical De-isolation Authorisation: |
|---|---|
| I Certify that the work has been completed and that I am one of the Acceptors of this Permit to Work: | I have been round and viewed the work area and am satisfied the work is complete.<br>Please de-isolate the items listed in 4b. |
| Signed:                              Print: | Signed:                              Print: |
| Position:                            Time: | (Authorised Person)               Time: |
| Company:                            Date: | Company:                            Date: |

| 12 - Equipment De-isolated (Electrical): | 13: Permit Cancelled:  (Authorised Person) |
|---|---|
| Signed:                              Print: | Signed:                              Print: |
| Position:                            Time: | (Authorised Person)               Time: |
| Company:                            Date: | Company:                            Date: |

# Clearance Certificate

Some companies use Clearance Certificates and some just rely on the Permit system. Let us take a motor and pump unit, a Mechanical Technician is going to work on the pump and pipework. If they are going to remove flange bolts on the pipeline then they would need to know that the chemical had been drained first. They would also need to know that the electric motor is not going to start. On some pipework systems, especially on the BP chemical factory where I came from, there may be electric trace heating on the pipework, which would require electrical isolation, and lagging which would have to be removed before the work could start.

If a pump was to be removed from the plant and taken to the workshop for overhaul then it must have gone through an operation where it was not only drained but steamed out at a steaming out bay to ensure that no trace of chemical remained. Remember, the workshop is a non-hazardous area. We must not turn it into a hazardous area by taking a pump full of hazardous chemical into it.

When carrying out work on the system the Mechanical Technician needs to know that the electric motor and any electric trace heating has been isolated.

At BP there were several types of Permit. The nearest to a Clearance Certificate was a 'C' type Permit:

An **'A' type Permit:** Used for entry into confined spaces, hot-work, radiation etc.
(Plant Manager's Signature)

A **'B' type Permit:** Which was in fact a Permit to Work.
(Plant Process Supervisor's Signature)

A **'C' type Permit:** Non tearable and attached to equipment that had been steamed out ready to be taken to the workshop. **(Closest to a Clearance Certificate.)**
(Mechanical Supervisor's/Technician's Signature)

A **'D' type Permit:** This was used for ships mooring at the jetty.
(Jetty Supervisor's Signature)

An **'A' Type Pass:** Used for test equipment, battery operated equipment, vehicles etc. **(Closest to a Gas Free Certificate.)**

(Plant Process Supervisor's Signature)

If companies are not careful, there are so many documents and the system can be so complex that nobody will be able to do anything.

**Certificate Number:**
CC1976

# Clearance Certificate

HOTA

| 3 - Description of work to be carried out: |
| --- |
| |
| |
| |
| |

| 4a - Process Isolations: | Time: | Date: | Initial: |
| --- | --- | --- | --- |
| 1 - | | | |
| 2 - | | | |
| 3 - | | | |

**4b - Electrical Isolations:   (Electrical Technician)**

Please Isolate the items below:          Signed:                    (Authorised Person)

| Distribution Board Number: | Circuit Number: | Fuse Size: | Date of Isolation: | Time of Isolation: | Signed by Technician: | Company: |
| --- | --- | --- | --- | --- | --- | --- |
| 1 - | | | | | | |
| 2 - | | | | | | |
| 3 - | | | | | | |

| 5 - Hazards Remaining: | 6 - Precautions to be taken: |
| --- | --- |
| 1 | 1 |
| 2 | 2 |
| 3 | 3 |

| 7 - Certification:  (Authorised Person) | 8 - Acceptance: |
| --- | --- |
| I Certify that the Isolations in Sections 4a & 4b have been carried out and the work may start subject to requirements of sections 5 & 6: | I have read and understand this Permit to Work and accept all the Hazards Remaining and the precautions to be taken: |
| Signed:                         Print: | Signed:                       Print: |
| (Authorised Person)             Time: | Position:                     Time: |
| Company:                        Date: | Company:                      Date: |

| 9 - Sign onto the permit (1 - 3): | | 9 - Sign onto the permit (4 - 6): | |
| --- | --- | --- | --- |
| 1 | Date: | 4 | Date: |
| 2 | Date: | 5 | Date: |
| 3 | Date: | 6 | Date: |

| 10 - Work Completed:  (Acceptor) | 11 - Electrical De-isolation Authorisation: |
| --- | --- |
| I Certify that the work has been completed and that I am one of the Acceptors of this Permit to Work: | I have been round and viewed the work area and am satisfied the work is complete. Please de-isolate the items listed in 4b. |
| Signed:                         Print: | Signed:                       Print: |
| Position:                       Time: | (Authorised Person)           Time: |
| Company:                        Date: | Company:                      Date: |

| 12 - Equipment De-isolated (Electrical): | 13: Permit Cancelled:  (Authorised Person) |
| --- | --- |
| Signed:                         Print: | Signed:                       Print: |
| Position:                       Time: | (Authorised Person)           Time: |
| Company:                        Date: | Company:                      Date: |

# Gas Free Certificate

A Gas Free Certificate may not be all that it sounds. The area that it covers is only gas free as long as your gas alarm or the plant evacuation alarm does not go off.

Remember this Certificate is given for work on a running plant, in most cases that would be Zone 2 - gas is unlikely or for short periods of time, as with Zone 1 the gas or vapour may be present in normal operation. In the case of Zone 1 a more stringent Permit system may be in place.

 There are several designs of gas monitor which would accompany the Gas Free Certificate. The most common alarm is similar to the design on the left which would fit into a technician's top pocket.

These alarm units are battery operated, but they would be Exia or Exib llC T6 which is the very best certification and would be suitable for any Zone, any gas group and T6 Temperature Classification i.e. any ignition temperature of any gas.

These units are failsafe so if the battery goes down it will go off.

Gas Free Certificates are given, usually by the Plant Process Supervisor if:

1  You are taking battery operated equipment, especially if it is not intrinsically safe, onto the plant. Some test instruments are Exia llC T6 which of course would not require a Gas Free Certificate.

2  Electrical Technicians are opening covers and testing for correct isolation. Supplies are considered to be live until proven to be dead.

3  A vehicle is being taken onto the plant. Diesel vehicles are usually no problem, but you will have more of a fight on your hands with a petrol vehicle because of the spark potential.

4  You are taking some non-intrinsically safe test equipment which may have its own generation unit. Older equipment, such as a winding Insulation Tester (Megger).

5  You are taking a torch or such equipment which is 'non-certified'.

6  You are taking electrical test equipment which could produce a voltage of 500v from its terminals.

7  Equipment with its own generation unit is being taken onto the plant e.g. a welding generator, including those with a small 110v generator on the side.

| Cert. Number:<br>1254G | # Gas Free Certificate |  |

**This Certificate is issued for:**

1 - Vehicles entering the Plant Area. (Diesel Vehicles are preferred!)
2 - Battery Operated Test Equipment.
3 - Non-Intrinsically Safe Test equipment
4 - Opening Electrical equipment where there may be a potentially live supply.

**1 - A Gas Monitor will be supplied with this Certificate, if it was to alarm then you must make your job safe and go to the muster point which in this case is the Control Room.**
**2 - If the Fire Alarm sounds make your job safe and go to the Muster Point which in this case is on the Green Lattice Area North of the Control Room.**

| **Type of Gas Monitor:** | | | | | |
|---|---|---|---|---|---|
| Make of Monitor: | | Type of Monitor: | | Number: | |

| **1 - Vehicles:** | | | | | |
|---|---|---|---|---|---|
| Make of Vehicle: | | Diesel / Petrol: | | Registration: | |
| Make of Vehicle: | | Diesel / Petrol: | | Registration: | |
| Make of Vehicle: | | Diesel / Petrol: | | Registration: | |

| **2 - Battery operated Test Equipment:** | | |
|---|---|---|
| What is the Item: | | Make & Model of Item: |
| What is the Item: | | Make & Model of Item: |
| What is the Item: | | Make & Model of item: |

| **3 - Non-Intrinsically Safe Test Equipment:** | | |
|---|---|---|
| What is the Item: | | Make & Model of Item: |
| What is the Item: | | Make & Model of Item: |
| What is the Item: | | Make & Model of item: |

| **4 - Opening Electrical Equipment where there is a potentially live supply:** | | |
|---|---|---|
| What are the Items: | | System: (i.e. Lighting) |

| **5 - Permit Number to Cross Reference: (If Applicable)** | | |
|---|---|---|
| Permit Number: | | Date of issue: |

| **6 - Certification: (Authorised Person)** | | **7 - Acceptance:** | |
|---|---|---|---|
| I have been out and viewed the work area and | | I will work fully within the scope of this Gas Free | |
| in my opinion it is safe for the above work to | | Certificate and will cease work in the event of | |
| continue. | | any alarms or requests from plant personnel. | |
| Signed: | Print: | Signed: | Print: |
| (Authorised Person) | Time: | Position: | Time: |
| Company: | Date: | Company: | Date: |

# Method Statement
## (Safe System of Work)

As with any task carried out on a chemical complex or platform, safety is the top priority both for personnel and plant. Method Statements explain procedures and how to achieve a particular task safely and in line with company policy.

We mentioned earlier that the Inspector(s) would need to know the scope of work and how far they are to go. It also must be remembered that the Inspector(s) might be contracted in from the outside so would need to firstly go through an Induction Session so that they know things like:

1   What the Fire Alarm sounds like and where the Fire Muster points are?

2   What the Plant Evacuation Alarm sounds like, how different is it from the Fire Alarm and where are the Muster Points here?

3   Where are the Safety Shower and Eye Wash Stations on the plant?

4   Where are the Fire Extinguisher Stations on the plant?

5   What Permit to Work System does the Company have, and where and who should they see to obtain a Permit in the first place?

6   What happens to the Permit to Work at the end of the day if the Task is not finished, and how is it updated?

7   Where do they report to each morning when they go to their place of work?

8   What if they see something that is unsafe, who do they see to get it rectified?

9   What Safety Equipment do they have to wear when they are on their place of work?

10   They are entitled by DSEAR to know what chemicals they are working with. How do they find this information out?

Above are several induction points that Inspectors need to be made aware of, I am sure that in some companies there will be more. Some companies will give out an Induction Procedure in written form as a matter of course.

On the next page I have drafted out what I consider to be a Method Statement, I am sure that some companies could add to this. It does point out the exact location of the work and the procedures that are required, including a Permit to Work.

This will go a long way towards keeping the Inspectors aware of their scope of work and what safety procedures are present.

# Method Statement for Detailed Inspection

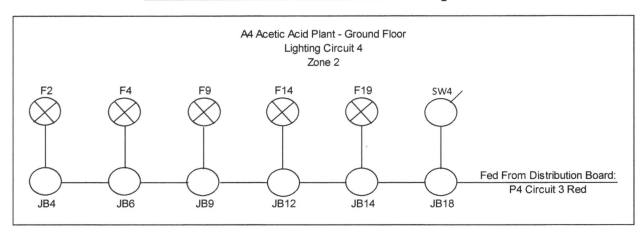

## Circuit Schematic Diagram

1 - Before completing the work, report to the plant control room and inform them of the scope of work to be done and explain that some lights will be out on the ground floor.

(I have given a lighting circuit as an example, the actual task could be anything the company required.)

2 - Obtain a Permit to Work.

3 - Please complete a **'Detailed Inspection'** to report form number D109 of the above lighting circuit consisting of 6 junction boxes, 5 bulkhead lights and a switch.

4 - The Inspection must be completed with the system isolated as there will be opening of covers into the enclosures.

5 - Do not include any lights, junction boxes or switches that are not on the above block diagram.

6 - Switch the lights on to the above circuit leaving the rest of the ground floor lighting in the off position. Then isolate the circuit ensuring the lights go out. **Complete a Full Isolation Procedure.**

7 - Hang a board on the switch **'Men Working on Lighting Circuit'** just to let anyone else know that the lights will be switched off.

8 - Carry out the inspection and complete the report.

9 – If you find anything that could be a danger to the plant such as a broken fitting, inform higher authority straight away.

10 – De-isolate the system and return the ground floor lighting to the position that it was when the work started.

11 - Return to the control room and inform them that the work is completed and sign off the Permit to Work.

I have read and understood the above Method Statement and will follow the procedure:

Name (Print):-                                      Signed:-                              Date:-

# Risk Assessment:
## Example

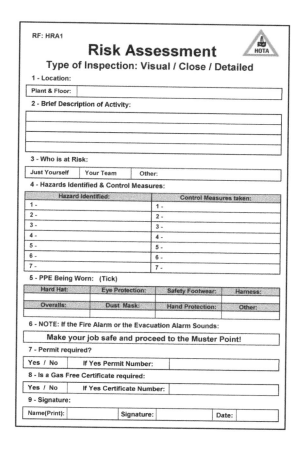

In industry the top priority is Safety to both people and plant. Urging people to carry out a Risk Assessment before they start the work goes a long way towards this objective and is a huge step in the right direction.

So how do we go about completing a Risk Assessment and what actually is it? A Risk Assessment is usually an assessment made by the person carrying out the work of all the dangers, large or small, that could harm either themselves or others in the vicinity and what precautions are to be taken to minimise these risks (dangers).

Some companies leave it to the individual to come up with some small report on which they have carried out their Assessment. I find that the easiest way is to draw up a simple form like the example on the left (full size copy on the next page) as a shell on a PC, and to run off a hard copy when required.

Do not make the form too complex as sometimes this deters a person from completing a job thoroughly. The Risk Assessment process is built into all of us and we use it to go through life every day especially when driving. If you are going to overtake another vehicle you actually carry out a Risk Assessment in your head as to whether or not you can overtake safely. If you think about it you probably do a very uncomplicated Risk Assessment in your head for most things that you do. So the process is there, built into us.

RF: HRA1

# Risk Assessment

## Type of Inspection: Visual / Close / Detailed

### 1 - Location:

| Plant & Floor: | |
|---|---|

### 2 - Brief Description of Activity:

| |
|---|
| |
| |
| |
| |

### 3 - Who is at Risk:

| Just Yourself | Your Team | Other: |
|---|---|---|

### 4 - Hazards Identified & Control Measures:

| Hazard Identified: | Control Measures taken: |
|---|---|
| 1 - | 1 - |
| 2 - | 2 - |
| 3 - | 3 - |
| 4 - | 4 - |
| 5 - | 5 - |
| 6 - | 6 - |
| 7 - | 7 - |

### 5 - PPE Being Worn: (Tick)

| Hard Hat: | Eye Protection: | Safety Footwear: | Harness: |
|---|---|---|---|
| | | | |
| Overalls: | Dust Mask: | Hand Protection: | Other: |
| | | | |

### 6 - NOTE: If the Fire Alarm or the Evacuation Alarm Sounds:

| Make your job safe and proceed to the Muster Point! |
|---|

### 7 - Permit required?

| Yes / No | If Yes Permit Number: | |
|---|---|---|

### 8 - Is a Gas Free Certificate required:

| Yes / No | If Yes Certificate Number: | |
|---|---|---|

### 9 - Signature:

| Name(Print): | | Signature: | | Date: | |
|---|---|---|---|---|---|

# Area Classification Drawing
## (Zones)

Before the Inspector(s) can identify whether the equipment they are inspecting is in the correct Zone they have to know what Zone they are in to start with. Back in the 60s & early 70s, chemical plants were divided into Divisions as in those days we copied the Americans who, I may add, are still in Divisions.

So plants were mostly Division 1 which, at the lower end, is similar to our Zone 1 areas. Remember the Americans do not have a Division 0 which is at the top end of Division 1 and is similar to our Zone 0. Division 2 equipment came in the 1970s.

In the UK there are three Zones for Gas and three Zones for Dust, namely:

1    **Zone 0:** This is an area where the Gas or Vapour is present **constantly** or for long periods of time:

2    **Zone 1:** This is an area where the Gas or Vapour is **likely** in normal operation.

3    **Zone 2:** This is an area where the Gas or Vapour is **not likely** in normal operation or for short periods of time.

With Dust things are slightly different. All we are interested in is a dust cloud as these can explode where layers burn. So just put a '2' in front of the Gas/Vapour Zone i.e. Zone **2** becomes Zone **22**.

1    **Zone 20:** This is an area where Dust in the form of a cloud is present **constantly** or for long periods of time.

2    **Zone 21:** This is an area where Dust in the form of a cloud is **likely** in normal operation.

3    **Zone 22:** This is an area where the Dust in the form of a cloud is **not likely** in normal operation or for short periods of time.

Many chemical plants these days are Zone 2 where the Gas or Vapour is not expected or only for short periods of time.

So Inspectors must have an **Area Classification Drawing** that will show which Zones the equipment is in that they are going to inspect.

**A4 Acetic Acid Plant**
**Ground Floor**
**Area Classification Drawing**

C104

C103

T101

V101

C102

V102

C101

Key
Zone 1
Zone 2

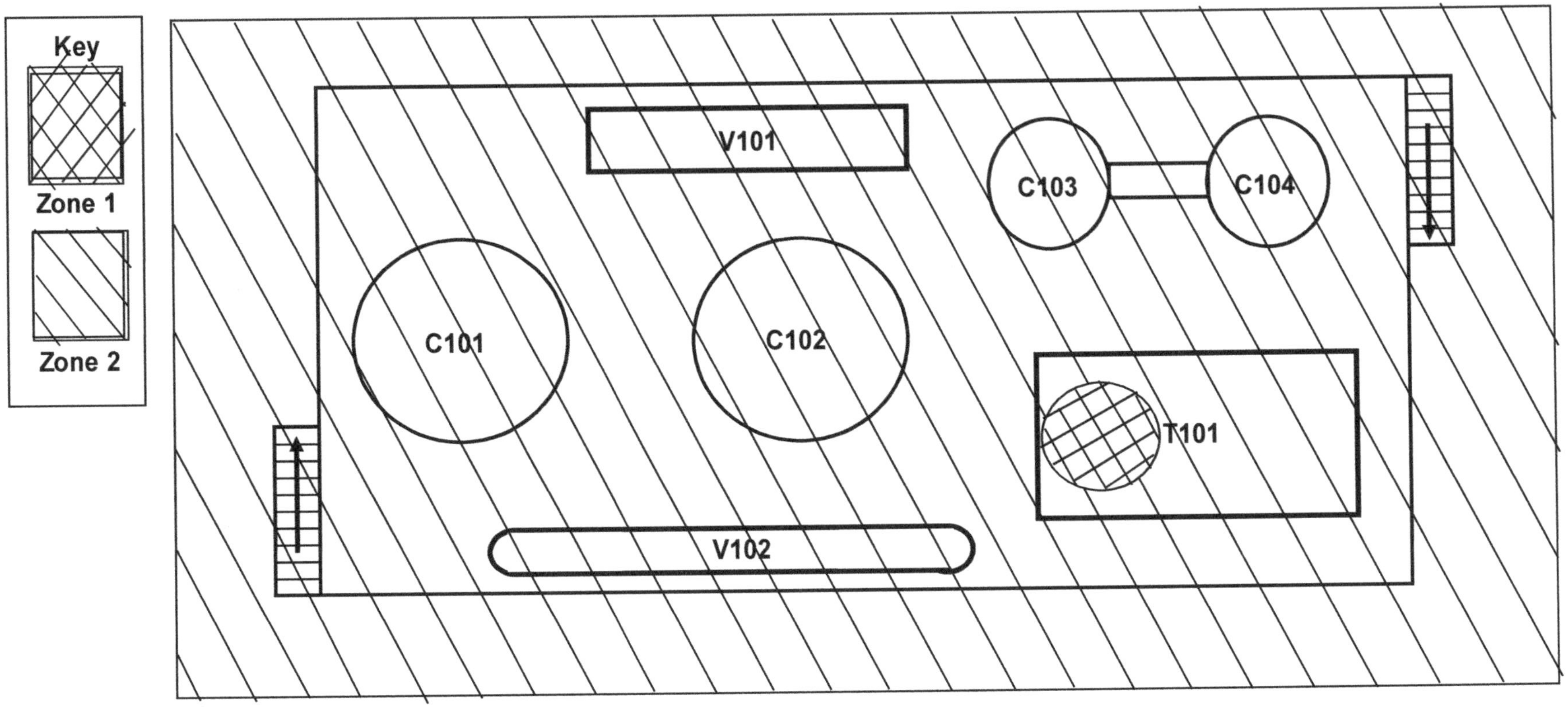

On a Visual Inspection you would have the above Area Classification Drawing

# Lighting Circuit Location Drawing
## (Fittings & Switches)

Again I have chosen a lighting circuit as an example of a typical Inspection.

On a plant they tend to mix the lighting circuits so that if a circuit fuse was to blow there are no huge dark spots on the plant.

Although the fittings should have identification numbers e.g. F1, F2 etc., remember where these lighting fittings are, as seeing the fitting number might be difficult and on a **Visual** Inspection there is no access equipment. Binoculars have been used before today on a Visual Inspection because of the 'no access' statement.

So the only way to get an idea of the fitting location is to have a location drawing of each floor and on this drawing would be all of the light fittings and their numbers along with light switches.

If an Inspector was instructed to inspect the lighting on a plant they would never do anything else so Sample Inspections could be the answer. At the company where I worked we did what were called 10% lighting inspections where we did 10% of the circuits one year and a different 10% the next year and in ten years we would have completed 100%.

In the meantime Electrical Technicians are changing faulty lamps so the fittings would get a type of detailed inspection there, and at night the process team just keep their eyes open for broken glasses etc.

The Inspector would take the lamp/circuit location drawing out of the master copy book and photostat it, putting the master copy back.

The Inspector would then take the Method Statement with the fitting locations, fitting numbers and switch location and numbers. and circle in red the fittings on the circuit to be inspected.

All of the floor lights should be switched off if possible and safe, leaving the circuit on that is to be inspected. A notice should be hung on the switch **'Men Working on the Lighting System'**.

The circuit should now be inspected to the spreadsheet list showing all of the points to be looked at along with the codes.

For a **Visual** and **Close** Inspection the lights can remain on as a reminder of the fittings to be inspected. This would also show up any faulty lamps.

On completion of the Inspection the lighting system should be returned to how it was before the Inspection and the notice removed from the switch.

# A4 Acetic Acid Plant
## Ground Floor
## Lighting Location Diagram

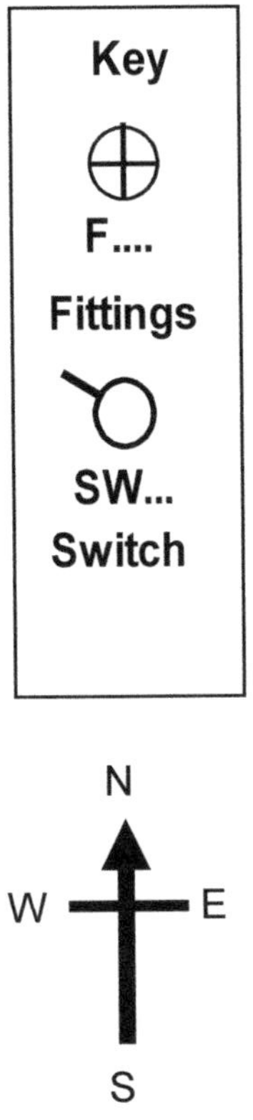

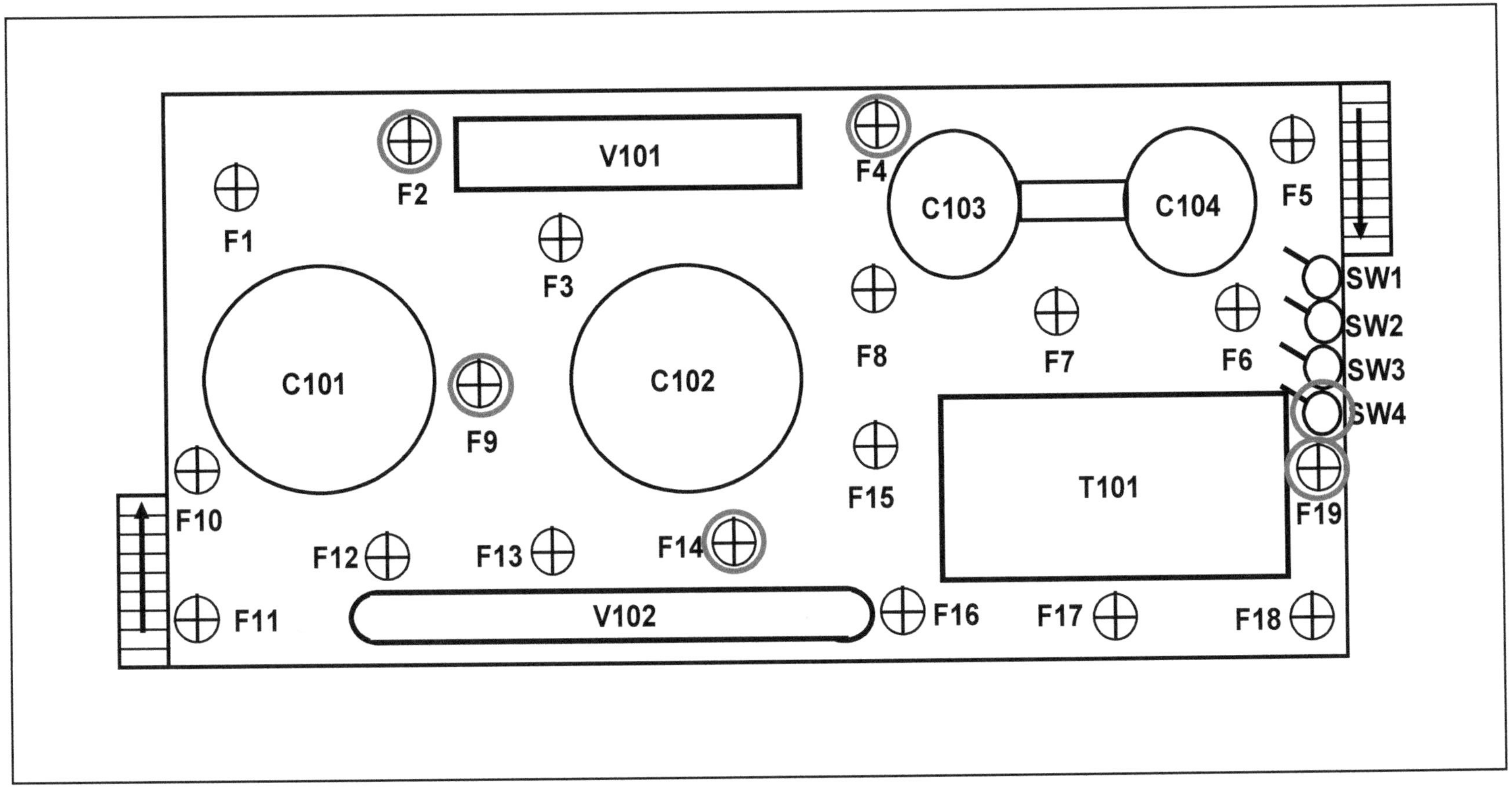

**Key**

⊕ F.... Fittings

◯ SW... Switch

W ← N | E / S

SW1 SW2 SW3 SW4

F5 F6 F7 F8 F4 F3 F2 F1

F19 F18 F17 F16 F15 F14 F13 F12 F11 F10 F9

C104 C103 C102 C101

V101 V102 T101

# Inspection Checklist

**Visual and Close Inspections** can be completed live and do not require any isolation. So unless you are using thermal image cameras, normal cameras, temperature guns etc. which by the way, can be used as there is no contact with the equipment, no Gas Free Certificate is required because there is no opening of any enclosures. This of course is subject to company policy.

**Detailed Inspections** are, however, a different matter as here equipment is going to be opened up to inspect the internal components. With Power Inspections (any voltage that is not intrinsically safe) then a full isolation of circuit has to be completed which means switching off circuit breakers or removing fuses and opening a junction box on that circuit in a Zone 2 area to check whether or not the correct supply has been isolated. Before this box is opened a Gas Free Certificate must be obtained because the tester may have got the wrong box and it could be live. **If the system is intrinsically safe then company policy applies to isolation.**

One of the documents is the Inspection Checklist and this lists the points the Inspector looks for when inspecting the equipment e.g. on a Visual Inspection: is the equipment in the correct Zone? On a Close Inspection: are glands tight? On a Detailed Inspection: are any wires loose on terminal blocks etc.?

Also with the above points, on Electrical and Instrument Inspections, are the codes that are in the IEC60079 Standard, that are applicable to each inspection, check? **This does not apply to Mechanical Inspections.** On the top of the Inspection Checklist is all the detail required as to location of the work to be done etc. Later in the book I have given an explanation and graphic descriptions of **my** interpretation as to the meaning of each point on the Checklist, why they are so important and why they must be done. **Remember if something is found to be dangerous it should be isolated until action is taken to remedy the problem.**

With this Checklist are other documents which may be required depending upon company policy. One of these documents is the **'Report Of Inspection'** which the person carrying out the Inspection completes for the engineer regarding the condition of the equipment and any problems that have arisen from the Inspection, the codes, and suggested remedial action.

A Permit to Work will always be required from the Plant Control Room Supervisor before any work can begin. Sometimes, depending upon company policy, other documents may be required before the Inspection can begin such as a Method Statement (safe system of work), a Risk Assessment etc.

Binoculars can be used without a Gas Free Certificate. Other items of equipment that are not intrinsically safe and have batteries in them such as thermal image cameras, temperature guns etc., will require a Gas Free Certificate and gas monitor. Gas Free Certificates, as mentioned above, are required for removing covers of junction boxes to check isolation.

# Inspection Report Checklist

## Protections: Exd - Exe - Exn - Ext

| Report Form Ref: | V157/A4 | Plant Ref: | A4 / Ground Floor |
|---|---|---|---|
| Inspection Type: | Visual | Zone: | Zone 2 |
| Equipment to be Inspected: | Plant Lighting Circuit 4 | | |
| Electrical Engineer: | I. Staff | Date: | |
| Name of Inspector: (Print:) | | Signed: | |

| Check: | Code: | Comments: |
|---|---|---|
| Is the Equipment in the correct Zone? | A1 | Zone 1 or Zone 2 |
| Is the Equipment degree of IP appropriate? | A5 | IP Washers |
| Is the Equipment Circuit Identification available? | A7 | |
| Is the Glass to Metal Sealing satisfactory? | A8 | |
| Is there any evidence of Unauthorised Modification? | A10 | |
| Are all of the Cover Bolts present? | | Including Grub Screws |
| Are all of the Cover Bolts Manufacturers? | A11 | |
| Are all of the Stoppers Present? | A11 | |
| Are all of the Stoppers correct Protection? | A1 | Univer… Exd or Exe |
| Are all Glands complete? | A11 | |
| Are the Glands the correct Protection? | A11 | …iversal, Exd or Exe |
| Are all Screw Covers undamaged? | A | Exd Equipment Only! |
| Are Fluorescent Tubes showi… End of Life …ts? | A26 | Fluorescent Lighting Only! |
| Are HID Lamps showing E… of Life Effects? | A27 | HID Lighting Only! |
| Do Motor Fans have s… cient cleara… e … the Cow… | A29 | Motors Only! |
| Is the Ventilation A… ow impeded? | A30 | Motors Only! |
| Is there any ob… us dama… to Cab… s? | B2 | |
| Are all Trunkin… Conduits, …cts … Pipes … ied satisfactorily | B3 | |
| Are Earth Bonds pr… nt? | B6 | |
| Are Earth Bonds correc… ss Sect… al Area? | B6 | 4mm2 |
| Are there any Obstructions a… t to Flanges? (Exd Only!) | B12 | Exd Equipment Only! |
| Is the Equipment adequately protected against Vibration? | C1 | |
| Is the Equipment adequately protected against the Weather? | C1 | |
| Is the Equipment adequately protected against Corrosion? | C1 | |
| Is the Equipment adequately protected against Temperature? | C1 | |
| Is there no undue accumulation of Dust or Dirt | C2 | |
| Is the Equipment visibly undamaged? | E1 | Not in the Standard |
| Is the Equipment ID available? | E2 | Not in the Standard |
| Is the Equipment Hazardous Data Plate readable? | E3 | Not in the Standard |

# Report on Inspection

The **'Report on Inspection'** is what the Inspectors complete to log all faults and problems with the equipment. This report should contain the 'Location' of the fault e.g. F1 (light fitting F1), what the fault is e.g. fitting in the wrong Zone/Gas Group etc., what Code from the Standard is applicable to this fault and what suggested remedial action should be taken.

Of course, the final remedial action will be decided by the Electrical, Instrument or Mechanical Engineer in charge of the plant's electrical, instrument or mechanical systems. Just a point, with Mechanical Inspections the procedure is exactly the same except here of course there are no codes and it is the Mechanical Engineer who will have the final say.

When you complete an Inspection you must be as thorough as possible as:

1    It might be a few years before this equipment is looked at again.

2    You are completing the Inspection ultimately for an Electrical, Instrument or Mechanical Engineer who is back in his or her office and not present with you.

3    Much of the equipment is vital to the running of the plant.

4    Anything dangerous that you find can be isolated/put right before something serious happens where the safety of the plant is compromised.

## Report on Inspection:

| Report Form Ref: | D095/A4 | Plant Ref: | A4 / Ground Floor |
|---|---|---|---|
| Inspection Type: | Detailed | Zone: | Zone 2 |
| Equipment to be Inspected: | Plant Lighting Circuit 4 | | |
| Electrical Engineer: | I. Staff | Date: | |
| Name of Inspector (Print) | | Signed: | |

| Location: | Fault: | Code: | Suggested Remedial Action: |
|---|---|---|---|
| | | | |
| | | | |
| | | | |

As you can see from the diagram above, the form is very easy to complete and on a large inspection there may be several forms completed. A full copy of this form is on the next page.

# Report on Inspection:

| Report Form Ref: | D095/A4 | Plant Ref: | A4 / Ground Floor |
|---|---|---|---|
| Inspection Type: | Detailed | Zone: | Zone 2 |
| Equipment to be Inspected: | Plant Lighting Circuit 4 | | |
| Electrical Engineer: | I. Staff | Date: | |
| Name of Inspector (Print)· | | Signed: | |

| Location: | Fault: | Code: | Suggested Remedial Action: |
|---|---|---|---|
| | | | |
| | | | |
| | | | |
| | | | |
| | | | |
| | | | |
| | | | |
| | | | |
| | | | |
| | | | |
| | | | |
| | | | |
| | | | |
| | | | |
| | | | |
| | | | |
| | | | |
| | | | |
| | | | |
| | | | |
| | | | |
| | | | |
| | | | |
| | | | |
| | | | |

# Electrical Schematic Wiring Diagram

As I have chosen a lighting circuit for our inspections there are in this case two types of diagram. There is the Main Wiring Diagram which shows the installer where every core is wired to, and a Schematic Wiring Diagram, as the one shown, which shows which piece of equipment is wired to another in the circuit, but no detail such as wiring core locations.

Usually the circuit being inspected is well established and the lights work so the Main Wiring Diagram would be of no use at all on a Visual or Close Inspection as junction boxes are not opened up so there is no need to know the detail of how the circuit is wired. A Schematic Diagram is ideal.

If you are fault finding then you may require the Main Wiring Diagram, but even on this occasion a Schematic may be sufficient. If you are wiring the circuit for the first time then this wiring diagram is required.

The diagram below is a very simple example of a Schematic Wiring Diagram. Just a point to mention that no matter what the circuit is e.g. lighting, a motor starter etc., **the circuit is always drawn de-energised (e.g. relays, switches, contactors etc. will be drawn de-energised).**

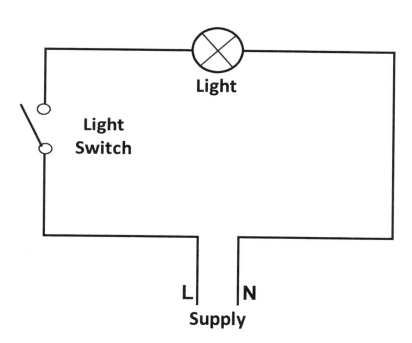

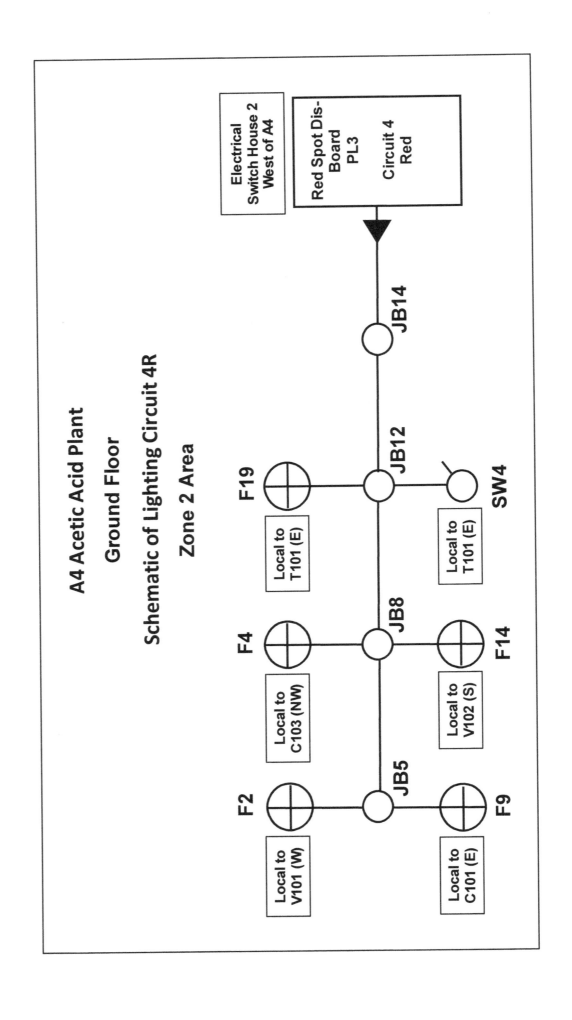

# Instrument Loop Diagram

An Instrument Loop Diagram is the equivalent of a Wiring Diagram in the electrical world. This diagram must be as near to 100% accurate as possible all of the time. This in fact is telling outside organisations such as the HSE **'This is how I am running my plant'** so if there is a visit by the HSE Inspector (Factory Inspector) and they want to have a look at a particular loop at random and the site is found to be different from the diagram then they would not be too pleased.

The Loop Diagram contains all of the detailed information on every item of equipment in the drawing such as barrier units, cables, instruments etc. so if anything is added, changed or uprated it is very important that this diagram is updated to incorporate the change.

So if you are looking, say, at a **'Close'** Inspection of instrumentation then all of the details on the instruments would be on this diagram for you to check against the site information on the instruments themselves. **Every detail should match.**

When the plant is built Loop Diagrams are produced for every instrument loop on the plant and these are what the construction company have got their instrument technicians to wire and install to. If any changes are made to the loop then some kind of company **'Mod. Form'** (Modification Form) should be produced and stapled to an archived drawing of what the loop was like when the plant was built so if the new modification does not turn out as expected then you will be able to see what it was like at every stage.

The detail of the equipment in the loop may look like below:

| Item: | Make: | Type Number: | Certification: | Notified Body: | Body Number: |
|---|---|---|---|---|---|
| Galvanic Barrier (1) | Pepperl & Fuchs | KFD2-CD2-Exi | II(1)G[Exia] IIC | BASEEFA | BAS00ATEX7240 |
| Galvanic Barrier (2) | MTL | MTL 3046B | [EExia] IIC | BASEEFA | Ex89C2112 |
| Zenner Barrier (1) | Pepperl & Fuchs | Z728F | II(1)GD[EExia] IIC | BASEEFA | BAS00ATEX7096 |
| Zenner Barrier (2) | Pepperl & Fuchs | Z728F | II(1)GD[EExia] IIC | BASEEFA | BAS00ATEX7096 |
| Junction Box JB1 | Weidmuller | TB1 | Exe II T6 | BASEEFA | BAS234 Ex7085 |
| Junction Box JB2 | Weidmuller | TB1 | Exe II T6 | BASEEFA | BAS234 Ex7085 |
| Barrier Box | Simple Apparatus | N/A | N/A | N/A | N/A |
| Field Device 1 | MTL | MTL892F | [EExia] IIC | BASEEFA | BAS01ATEX7012 |
| Field Device 2 | Pepperl & Fuchs | NCN15-MIK-NO | Eexia IIC T4 | PTB | PTB00ATEX2032X |

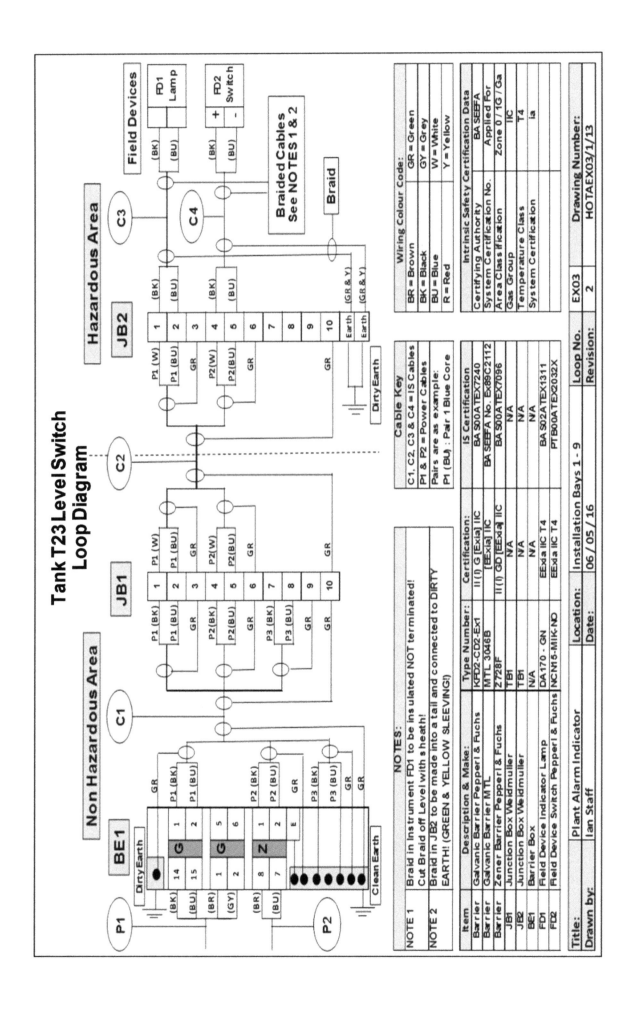

# Tank T23 Level Switch Loop Diagram

**Field Devices**

| FD1 Lamp |
| FD2 Switch |

**Hazardous Area**

**Non Hazardous Area**

## Wiring Colour Code:

| BR = Brown | GR = Green |
| BK = Black | GY = Grey |
| BU = Blue | W = White |
| R = Red | Y = Yellow |

## Cable Key

| C1, C2, C3 & C4 = IS Cables |
| P1 & P2 = Power Cables |
| Pairs are as example: |
| P1 (BU) : Pair 1 Blue Core |

## NOTES:

| NOTE 1 | Braid in Instrument FD1 to be insulated NOT terminated! Cut Braid off Level with sheath! |
| NOTE 2 | Braid in JB2 to be made into a tail and connected to DIRTY EARTH! (GREEN & YELLOW SLEEVING!) |

## Intrinsic Safety Certification Data

| Certifying Authority | BA SEFFA |
| System Certification No. | Applied For |
| Area Classification | Zone 0 / 1G / Ga |
| Gas Group | IIC |
| Temperature Class | T4 |
| System Certification | ia |

| Item | Description & Make: | Type Number: | Certification: | IS Certification |
|---|---|---|---|---|
| Barrier | Galvanic Barrier Pepperl & Fuchs | KFD2-CD2-Ex1 | II (I) G [Exia] IIC | BA S00ATEX7240 |
| Barrier | Galvanic Barrier MTL | MTL 3046B | [EExia] IIC | BA SEFFA No. Ex89C2112 |
| Barrier | Zener Barrier Pepperl & Fuchs | Z728F | II (I) GD [EExia] IIC | BA S00ATEX7096 |
| JB1 | Junction Box Weidmuller | TB1 | N/A | N/A |
| JB2 | Junction Box Weidmuller | TB1 | N/A | N/A |
| BE1 | Barrier Box | N/A | N/A | N/A |
| FD1 | Field Device Indicator Lamp | DA170 - GN | EExda IIC T4 | BA S02ATEX1311 |
| FD2 | Field Device Switch Pepperl & Fuchs | NCM15-MIK-NO | EExda IIC T4 | PTB00ATEX2032X |

| Title: | Plant Alarm Indicator | Location: | Installation Bays 1 - 9 | Loop No. | EX03 | Drawing Number: |
| Drawn by: | Ian Staff | Date: | 06 / 05 / 16 | Revision: | 2 | HOTAEX03/1/13 |

42

# Piping and Instrument Diagram (P & ID)

A **'Piping and Instrument Diagram'** (P & ID) is sometimes referred to as a 'process and instrument diagram'. This is similar to a Schematic Diagram, but shows all of the pipework, valves and instruments in a particular process system. The diagram also includes flow arrows for particular liquids, steam and gases in the system. A P & ID may be required on a Mechanical Inspection.

Another department that may find a P & ID handy is the Instrument Department. The P & ID will show where in the process all of the instrumentation and control valves are situated. Some equipment symbols you might come across are in the diagrams below:

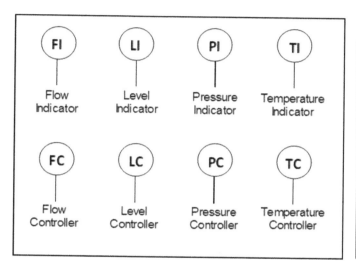 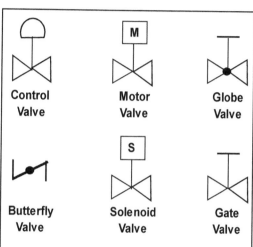

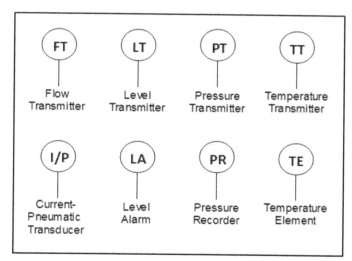

# R7 Production Plant

## Piping and Instrument Diagram (P & ID)

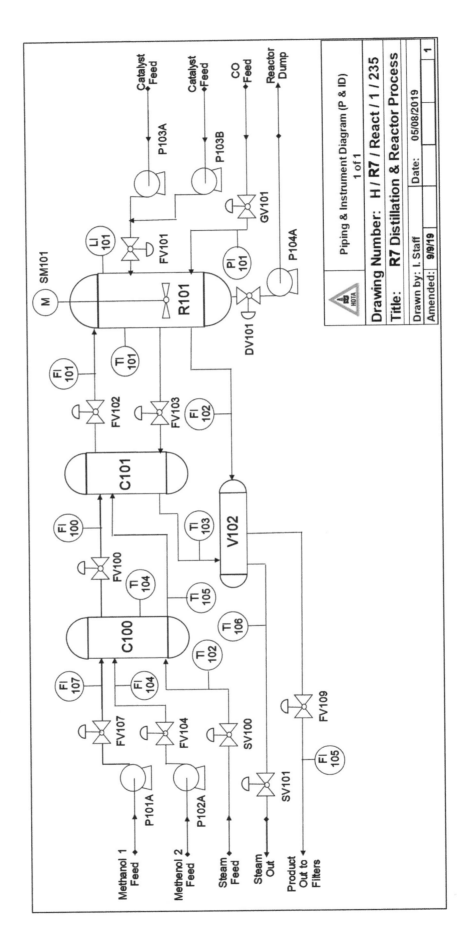

| | | |
|---|---|---|
| Piping & Instrument Diagram (P & ID) 1 of 1 | | |
| **Drawing Number:** | **H / R7 / React / 1 / 235** | |
| **Title:** | **R7 Distillation & Reactor Process** | |
| Drawn by: I. Staff | Date: 05/08/2019 | |
| Amended: 9/9/19 | | 1 |

# Visual Inspection

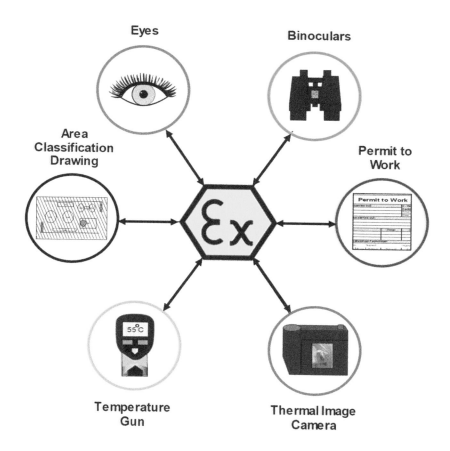

# Visual Inspection Description

Looking at the **Electrical** Inspection, just being '**Visual**' it can be completed live and does not require any isolation. So unless you are using thermal image cameras, normal cameras, temperature guns etc. which, by the way, can be used as there is no contact with the equipment, no Gas Free Certificate is required because there is no opening of any enclosures. (It may be company policy.)

One of the documents is the Inspection Checklist and this lists the points the Inspector looks for when inspecting the equipment, e.g. is the equipment in the correct Zone for its protection, are there any bolts, screws or stoppers missing etc. This list is obtained from the IEC60079 Standard, but I have produced a simpler version in the next pages. Also with the above points, on **Electrical** and **Instrument** Inspections, are the Codes that are in the IEC60079 Standard that are applicable to each inspection check. **Codes do not apply to Mechanical Inspections.** On the top of the Inspection Checklist is all the detail required as to location of the work to be done.

Later in the book I have given an explanation and graphic descriptions of **my interpretation** as to the meaning of each point on the Checklist, why they are so important and why they must be done. With this Checklist are other documents that may be required, depending upon company policy. One of these documents is the Report of Inspection, which the person carrying out the Inspection completes for the engineer regarding condition of the equipment and any problems that arose from the Inspection, the Codes and suggested remedial action.

The manual states that on an **Electrical** or **Instrument** 'Visual' Inspection the Inspector would have just an Area Classification Drawing, no access equipment such as ladders or steps, no tools and no wiring diagrams, schematic diagrams or loop diagrams in the case of instrumentation. A Permit to Work will always be required from the Plant Control Room Supervisor before any work can begin. Sometimes depending upon company policy other documents may be required before the Inspection can begin such as a Method Statement (safe system of work), a Risk Assessment etc.

Items that can be used without a Gas Free Certificate are binoculars. Other items of equipment that are not intrinsically safe and have batteries in them such as thermal image cameras, temperature guns etc., will require a Gas Free Certificate and gas monitor. **Electrical** or **Instrument** items from the Standard are marked **A, B,** or **C,** but you will see on the bottom of the Checklist that there are several marked '**E' (E stands for extra).** These are faults which, **in my opinion,** should be looked for, but are not in the Standard. If you look at the Checklists on the next few pages you will see what I mean.

A **Mechanical 'Visual'** Inspection is slightly different, here the Inspector would have an '**Area Classification Drawing'** of the plant or platform, and a '**P & ID'** of the system to be worked upon. The P & ID would be equivalent to an electrical or instrument '**Schematic'** and a '**Location Diagram'** and this is a drawing of the pipework and instrumentation. There is no such thing as a **Mechanical 'Close'** Inspection so Inspectors can feel for loose bolts etc. on the '**Visual'** Inspection.

# Visual Inspections
## Documentation Required

In all of the literature on **'Visual'** Inspections it says that the only document allowed/required is an Area Classification Drawing. On other Inspections Schematic/Loop Diagrams are required. Below are several extra documents that may be required on various Inspections depending upon company procedures and policies. Depending upon which Inspection is being carried out, as per title at the top, the documents that may be required on a **'Visual'** Inspection are highlighted below in Red:

1   Permit to Work: This depends solely on company policy, but the chances are that the Inspector(s) would not be allowed on the plant without one.

2   **A Clearance Certificate:** Some companies do not use or rely on these Clearance Certificates and just use a Permit to Work.

3   **A Gas free Certificate:** Used when battery & non-IS test equipment is being used or when covers are being opened where there is a potentially live supply. Required on an Electrical Detailed Inspection.

4   A Method Statement: Written by the Electrical, Instrument or Mechanical Engineer, stating the safe system of work and will also include the scope of work. (e.g. How far do the Inspectors go?)

5   A Risk Assessment: The Permit will go a long way towards this, but that is a company document whereas this Assessment is completed by the Inspector(s).

6   An Area Classification Drawing: The Inspector(s) would need to know what Zone they are in so they can ensure that the equipment is suitable.

7   An Equipment Location Diagram: The Inspectors(s) would need to know exactly where on the plant the equipment is that they have to inspect.

8   The Checklist: This would show what faults they would be looking for on an inspection and what codes would go with those faults.

9   The Report: This is the form where the Inspector(s) write down the location, fault, code and suggested remedial action.

10  **A Circuit Wiring/Schematic Diagram:** If the equipment is a power circuit such as lighting, motors, sockets etc. Usually a Schematic Diagram is used on an Inspection, a Wiring Diagram would not be a great deal of use here and would be more use on an Installation Project.

11  **A Loop Diagram:** If the equipment is instrumentation the loop diagram would be invaluable as all the information on the instruments and barrier units would be on here. This drawing should be as near 100% accurate as possible all of the time.

12  **A Piping and Instrumentation Diagram (P & ID):** This would be required on the Mechanical Inspection to show the pipework included in the Inspection.

# Inspection Report Checklist
## Protections Exd, Exe, Exn & Ext

| Report Form Reference: | V157 / A4 | Plant Reference: | A4 / Glound Floor |
|---|---|---|---|
| Inspection Grade: | Visual | Zone: | Zone 2 |
| Equipment to be Inspected: | Plant Lighting Circuit 4 from Dis-Board PL2 | | |
| Responsible Electrical Engineer: | I. Staff | Date: | |
| Name of Inspector (Print): | | Signed: | |

| Check: | Code: | Comments: |
|---|---|---|
| Is the Equipment in the correct Zone? | A1 | Zone 1 or Zone 2 |
| Is the Equipment degree of IP appropriate? | A5 | IP Washers |
| Is the Equipment Circuit Identification available? | A7 | |
| Is the Glass to Metal Sealing satisfactory? | A8 | |
| Is there any evidence of Unauthorised Modifications? | A10 | |
| Are all of the Cover Bolts present? | A11 | Including Grub Screws |
| Are all of the Cover Bolts the Manufacturer's? | A11 | |
| Are all of the Stoppers present? | A11 | |
| Are all of the Stoppers the correct Protection? | A11 | Universal, Exd or Exe |
| Are all of the Glands complete? | A11 | |
| Are all of the Glands the correct Protection? | A11 | Universal, Exd or Exe |
| Are all of the Covers the correct type & undamaged? | A12 | Exd Equipment only |
| Are any Fluorescent Tubes displaying End of Life? | A26 | Fluorescent Lighting only |
| Are any HID Lamps displaying End of Life? | A27 | HID Lighting only |
| Do Motor Fans have sufficient clearance with the Cowl? | A29 | Motors only |
| Is the Motor Ventilation Air Flow impeded? | A30 | Motors only |
| Is there any obvious damage to Cables? | B2 | |
| Are all Trunkings, Conduits, Ducts & Pipes sealed satisfactorily? | B3 | |
| Are all Earth Bonds present? | B6 | |
| Are all Earth Bonds sufficient Cross Sectional Area? | B6 | 4mm2 |
| Are there any Obstructions adjacent to Flanges? | B12 | Exd Equipment only |
| Is the Equipment adequately protected against Vibration? | C1 | |
| Is the Equipment adequately protected against the Weather? | C1 | |
| Is the Equipment adequately protected against the Corrosion? | C1 | |
| Is the Equipment adequately protected against the Temperature? | C1 | |
| Is there any accumulation of Dust or Dirt? | C2 | |
| Is the outside of the Equipment undamaged? | E1 | Not in the Standard |
| Is the Equipment ID available? | E2 | Not in the Standard |

# Inspection Report Checklist
# Protections Exi

| Report Form Reference: | V127 / A4 | Plant Reference: | A4 / 1st Floor |
|---|---|---|---|
| Inspection Grade: | Visual | Zone: | Zone 2 |
| Equipment to be Inspected: | | Acetic Acid Tank (T23) Level loop: IS / 22 / Lev.DP | |
| Responsible Electrical Engineer: | I. Staff | Date: | |
| Name of Inspector (Print): | | Signed: | |

| Check: | Code: | Comments: |
|---|---|---|
| Is the Equipment in the correct Zone? | A1 | Zone 1 or Zone 2 |
| Is the Equipment Documentation appropriate for the Zone? | A1 | Zone 1 or Zone 2 |
| Are all of the Stoppers present? | A10 | |
| Are all of the Stoppers the correct Protection? | A10 | Universal, Exd or Exe |
| Are all of the Glands complete? | A10 | |
| Are all of the Glands the correct Protection? | A10 | Universal, Exd or Exe |
| Are all of the Glands the correct Type? | A10 | Compression or SWA |
| Are all Safety Barriers the correct way round? | A13 | Barrier Units |
| Are all Zener Barriers Earthed? | A13 | Barrier Units |
| Are all Zener Barrier Earths connected to Clean Earth? | A13 | Barrier Units |
| Is the Barrier Enclosure Clean Earth Feed Wire 4mm2 | A13 | Barrier Units |
| Is the Barrier Enclosure Dirty Earth Feed Wire 4mm2 | A13 | Barrier Units |
| Are Zener and Galvanic Barriers on separate rails? | A13 | Barrier Units |
| Is there any damage to Cables? | B3 | |
| Are all Trunkings, Conduits, Ducts & Pipes sealed satisfactorily? | B4 | |
| Is the Equipment adequately protected against Vibration? | C1 | |
| Is the Equipment adequately protected against the Weather? | C1 | |
| Is the Equipment adequately protected against Corrosion? | C1 | |
| Is the Equipment adequately protected against the Temperature? | C1 | |
| Is there any accumulation of Dust or Dirt? | C2 | |
| Have JBs with just one Loop got a Blue Label? | E1 | Not in the Standard |
| Are Junction Boxes with Multi-Loops Certified? | E2 | Not in the Standard |
| Is the Equipment Circuit ID available? | E3 | Not in the Standard |
| Are all the Cover Bolts present? | E4 | Not in the Standard |
| Is the Equipment ID available? | E5 | Not in the Standard |
| Is there any obvious Damage to Equipment? | E6 | Not in the Standard |
| Is the Equipment Hazardous Data Plate readable? | E7 | Not in the Standard |

# Inspection Report Checklist
## Protections Exp

| Report Form Reference: | V27 / A4 | Plant Reference: | A4 / Ground Floor |
|---|---|---|---|
| Inspection Grade: | Visual | Zone: | Zone 2 |

| Equipment to be Inspected: | Instrument Pressurised Cabinet A4/P23 |
|---|---|

| Responsible Electrical Engineer: | I. Staff | Date: | |
|---|---|---|---|
| Name of Inspector (Print): | | Signed: | |

| Check: | Code: | Comments: |
|---|---|---|
| Is the Equipment in the correct Zone? | A1 | Zone 1 or Zone 2 |
| Is the Equipment Cable/Circuit ID available? | A5 | |
| Are any Glass to Metal Seals intact? | A6 | |
| Is there any evidence of Unauthorised Modifications? | A8 | |
| is there any damage to Cables? | B2 | |
| Is the Earth Bonding Wire connected to Earth? | B3 | |
| Is the Earth Bonding Wire 4mm2? | B3 | |
| Are the Pressurisation Ducts and Pipes in good condition? | B8 | |
| Is the Protective Gas free from Contaminates? | B9 | |
| Is the Protective Gas Pressure, or Flow, adequate? | B10 | |
| Is the Equipment protected against the Weather? | C1 | |
| Is the Equipment protected against Corrosion? | C1 | |
| Is the Equipment protected against Vibration? | C1 | |
| Is the Equipment protected against Temperature? | C1 | |
| Is there any accumulation of Dust or Dirt? | C2 | |
| Is there any damage to the outside of the Equipment? | E1 | Not in the Standard |
| Is the Equipment ID available? | E2 | Not in the Standard |
| Are there any open holes in the equipment? | E3 | Not in the Standard |
| Do all glands look sealed into the Enclosure? | E4 | Not in the Standard |

# Report on Inspection:

| Report Form Ref: | VM217/A4 | Plant Ref: | A4 / Ground Floor |
|---|---|---|---|
| Inspection Type: | Visual | Zone: | Zone 2 |
| Equipment to be Inspected: | Plant Lighting Circuit 4 | | |
| Electrical Engineer: | I. Staff | Date: | |
| Name of Inspector (Print) | | Signed: | |

| Location: | Fault: | Code: | Suggested Remedial Action: |
|---|---|---|---|
| | | | |
| | | | |
| | | | |
| | | | |
| | | | |
| | | | |
| | | | |
| | | | |
| | | | |
| | | | |
| | | | |
| | | | |
| | | | |
| | | | |
| | | | |
| | | | |
| | | | |
| | | | |
| | | | |
| | | | |
| | | | |
| | | | |
| | | | |
| | | | |
| | | | |

# Visual Inspections (Mechanical)
## Documentation Required

In all of the literature on 'Visual' Inspections it says that the only document allowed/required is an Area Classification Drawing. On other Inspections Schematic/Loop Diagrams are required. Below are several extra documents that may be required on various Inspections depending upon company procedures and policies. Depending upon which Inspection is being carried out, as per title at the top, the documents that may be required on a 'Visual' Inspection are highlighted below in Red:

1   Permit to Work: This depends solely on company policy, but the chances are that the Inspector(s) would not be allowed on the plant without one.

2   A Clearance Certificate: Some companies do not use or rely on these Clearance Certificates and just use a Permit to Work.

3   **A Gas free Certificate:** Used when battery & non-IS test equipment is being used or when covers are being opened where there is a potentially live supply. Required on an Electrical Detailed Inspection.

4   A Method Statement: Written by the Electrical, Instrument or Mechanical Engineer stating the safe system of work and will also include the scope of work. (e.g. How far do the Inspectors go?)

5   A Risk Assessment: The Permit will go a long way towards this, but that is a company document and this Assessment is completed by the Inspector(s).

6   An Area Classification Drawing: The Inspector(s) would need to know what Zone they are in so they can ensure that the equipment is suitable.

7   An Equipment Location Diagram: The Inspectors(s) would need to know exactly where on the plant the equipment is that they have to inspect.

8   The Checklist: This would show what faults they would be looking for on an Inspection and what codes would go with those faults.

9   The Report: This is the form where the Inspector(s) write down the location, fault, code and suggested remedial action.

10  **A Circuit Wiring/Schematic Diagram:** If the equipment is a power circuit such as lighting, motors, sockets etc. Usually a Schematic Diagram is used on an Inspection, a Wiring Diagram would not be a great deal of use here and would be more use on an Installation Project.

11  **A Loop Diagram:** If the equipment is instrumentation the Loop Diagram would be invaluable as all the information on the instruments and barrier units would be on here. This drawing should be as near 100% accurate as possible all of the time.

12  A Piping and Instrumentation Diagram (P & ID): This would be required on the Mechanical Inspection to show the pipework included in the Inspection.

# Inspection Report Checklist
## Protection Mechanical

| Report Form Reference: | VM217/A4 | Plant Reference: | A4 / Ground Floor |
|---|---|---|---|
| Inspection Grade: | Visual | Zone: | Zone 2 |
| Equipment to be Inspected: | | Column C101 Feed Pump P10A | |
| Responsible Electrical Engineer: | I. Staff | Date: | |
| Name of Inspector (Print): | | Signed: | |

| Check: | Code: | Comments: |
|---|---|---|
| Is the Electric Motor Atex? | N/A | Electric Motor |
| Is the Motor in the correct Zone? | N/A | Electric Motor |
| Is the Motor the correct Gas group? | N/A | Electric Motor |
| Are there any unplugged holes in the Motor? | N/A | Electric Motor |
| Are there any problems with the Electric Cables or Glands? | N/A | Electric Motor |
| Is the Motor Casting damaged? | N/A | Electric Motor |
| Are there any Eye Bolts still in the Motor? | N/A | Electric Motor |
| Is the Motor purposely Earthed with an Earth wire? | N/A | Electric Motor |
| Is the Motor Fan Cowl damaged in any way? | N/A | Electric Motor |
| Is the Motor Air Flow restricted in any way? | N/A | Electric Motor |
| Are any Bolts missing from the Motor Covers? | N/A | Electric Motor |
| Does the direction of the Motor match the Pump? | N/A | Electric Motor |
| Is the Motor ID Number present and correct? | N/A | Electric Motor |
| Is the Pump Atex? | N/A | Pump |
| Are all Nuts present in the Pump Volute? | N/A | Pump |
| Is there any physical damage to the Pump Casting? | N/A | Pump |
| Are Constant Oilers full of Oil? | N/A | Pump |
| Is the Pump purposly Earthed with an Earth Wire? | N/A | Pump |
| Is the Pump ID present and correct? | N/A | Pump |
| Are any Valves in the System not Atex? | N/A | Valves |
| Are all Valve Flow Direction arrows correct? | N/A | Valves |
| Do all Flanges look as if a Gasket is present? | N/A | Flanges |
| Are all Flange Bonding Links in Good condition? | N/A | Flanges |
| Are any Stud Bolts loose? | N/A | Flanges |
| Are any Stud Bolts too short? | N/A | Flanges |
| Are any Stud Bolts too long? | N/A | Flanges |
| Are any Nuts on the Stud Bolts installed incorrectly? | N/A | Flanges |
| Is the Coupling Guard Atex? | N/A | Coupling Guard |
| Is the Coupling Guard the correct Metal? | N/A | Coupling Guard |
| Is there any damage to the Coupling Guard? | N/A | Coupling Guard |

# Report on Mechanical Inspection:

| Report Form Ref: | VM217/A4 | Plant Ref: | A4 / Ground Floor |
|---|---|---|---|
| Inspection Type: | Visual | Zone: | Zone 2 |
| Equipment to be Inspected: | Column C101 feed Pump P10A | | |
| Mechanical Engineer: | I. Staff | Date: | |
| Name of Inspector (Print): | | Signed: | |

| Location: | Fault: | Code: | Suggested Remedial Action: |
|---|---|---|---|
| | | N/A | |
| | | N/A | |
| | | N/A | |
| | | N/A | |
| | | N/A | |
| | | N/A | |
| | | N/A | |
| | | N/A | |
| | | N/A | |
| | | N/A | |
| | | N/A | |
| | | N/A | |
| | | N/A | |
| | | N/A | |
| | | N/A | |
| | | N/A | |
| | | N/A | |
| | | N/A | |
| | | N/A | |
| | | N/A | |
| | | N/A | |
| | | N/A | |
| | | N/A | |

# Close Inspection

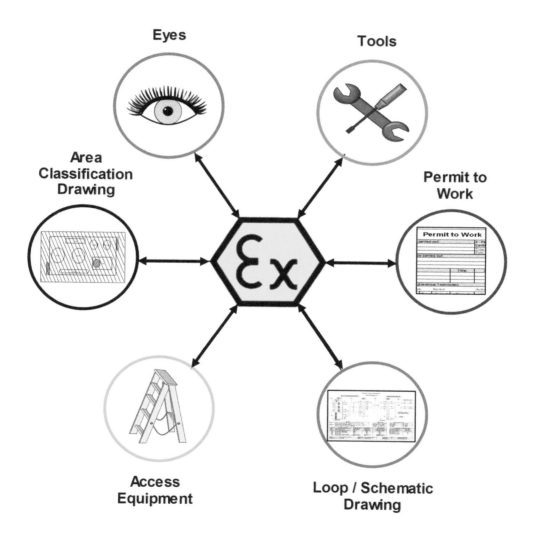

# Inspection Checklist Description

The Inspection Checklist lists the points that the Inspector looks for when inspecting the equipment, e.g. are the glands, bolts, lids, screws and stoppers tight? Also with these points are the Codes that are in the IEC60079 Standard that are applicable to each Inspection Check. On the top of the Inspection Checklist are all of the details required as to location of the work to be done etc.

Later in the book I have given an explanation and graphic description of **my interpretation** as to the meaning of each point on the Checklist and why it must be done.

With this document, comes, depending upon company policy, the Area Classification Drawing, the Schematic or Loop Diagram, the Method Statement and the Equipment Location Diagram. The Inspector is now ready to inspect the equipment.

The Report is completed by the Inspector for the engineer, regarding the condition of the equipment, any problems that arose from the Inspection, and suggested remedial actions, although in the end it is the engineer who will decide what is to be done and when.

Remember this is a '**Close**' Inspection so as well as more documentation, access equipment and tools can be used only to check tightness, **not to open up.** Other items that can be used are binoculars, thermal image cameras and temperature guns, although these should have been used on the Visual Inspection.

# Close Inspections
## Documentation Required

In all of the literature on **'Visual'** Inspections it says that the only document allowed/required is an Area Classification Drawing. On a **'Close'** Inspection Schematic/Loop Diagrams are required. Below are several extra documents that may be required on various Inspections depending upon company procedures and policies. Depending upon which Inspection is being carried out, as per title at the top, the documents that may be required on a **'Close'** Inspection are highlighted below in Red:

1. Permit to Work: This depends solely on company policy, but the chances are that the Inspector(s) would not be allowed on the plant without one.

2. **A Clearance Certificate:** Some Companies do not use or rely on these Clearance Certificates and just use a Permit to Work.

3. **A Gas Free Certificate:** Used when battery & non-IS test equipment is being used or when covers are being opened where there is a potentially live supply. Required on an Electrical Detailed Inspection.

4. A Method Statement: Written by the Electrical, Instrument or Mechanical Engineer stating the safe system of work and will also include the scope of work. (e.g. How far do the Inspectors go?)

5. A Risk Assessment: The Permit will go a long way towards this, but that is a company document whereas this Assessment is completed by the Inspector(s).

6. An Area Classification Drawing: The Inspector(s) would need to know what Zone they are in so they can ensure that the equipment is suitable.

7. An Equipment Location Diagram: The Inspectors(s) would need to know exactly where on the plant the equipment is that they have to inspect.

8. The Checklist: This would show what faults they would be looking for on an Inspection and what codes would go with those faults.

9. The Report: This is the form where the Inspector(s) write down the location, fault, code and suggested remedial action.

10. A Circuit Wiring/Schematic Diagram: If the equipment is a power circuit such as lighting, motors, sockets etc. Usually a Schematic Diagram is used on an Inspection, a Wiring Diagram would not be a great deal of use here and would be more use on an Installation Project.

11. A Loop Diagram: If the equipment is instrumentation the Loop Diagram would be invaluable as all the information on the instruments and barrier units would be on here. This drawing should be as near 100% accurate as possible all of the time.

12. **A Piping and Instrumentation Diagram (P & ID):** This would be required on the Mechanical Inspection to show the pipework included in the Inspection.

# Inspection Report Checklist
# Protections Exd, Exe, Exn & Ext

| Report Form Reference: | C166 / A4 | Plant Reference: | A4 / Glound Floor |
|---|---|---|---|
| Inspection Grade: | Close | Zone: | Zone 2 |
| Equipment to be Inspected: | Plant Lighting Circuit 4 from Dis-Board PL2 | | |
| Responsible Electrical Engineer: | I. Staff | Date: | |
| Name of Inspector (Print): | | Signed: | |

| Check: | Code: | Comments: |
|---|---|---|
| Is the Equipment Group appropriate? | A2 | Group I or II + Gas Group |
| Is the Equipment Temperature Class appropriate? | A3 | T1 - T6 |
| Is the Equipment Maximum Surface Temperature correct? | A4 | Dust only |
| Are all of the Cover Bolts tight? | A11 | Including Grub Screws |
| Are all of the Stoppers tight? | A11 | |
| Is a Standard Gland required or a Barrier Gland? | A11 | |
| Are all of the Glands tight? | A11 | |
| Are all of the Screw Covers tight? | A12 | Exd Flameproof only |
| Is the Flameproof Gap within tolerance? | A16 | Exd Flameproof only |
| Are Breathing and Draining Accessories correct Protection? | A25 | |
| Is the Earth Bond Wire tight? | B6 | Physical Check |
| Is Variable Voltage / Frequency Equipment as Documentation? | B13 | |
| Is the Equipment ID correct? | E3 | Not in the Standard |
| Are the Electric Motor Bearings satisfactory? (SPM etc Test) | E4 | Not in the Standard |

# Inspection Report Checklist
## Protections Exi

| Report Form Reference: | C126 / A4 | Plant Reference: | A4 / 1st Floor |
|---|---|---|---|
| Inspection Grade: | Close | Zone: | Zone 2 |

| Equipment to be Inspected: | Acetic Acid Tank (T23) Level loop: IS / 22 / Lev.DP |
|---|---|

| Responsible Electrical Engineer: | I. Staff | Date: | |
|---|---|---|---|
| Name of Inspector (Print): | | Signed: | |

| Check: | Code: | Comments: |
|---|---|---|
| Is the Equipment that specified in the Documentation? | A2 | Loop Diagram etc. |
| Are the Barrier Units those specified in the Documentation? | A2 | Loop Diagram etc. |
| Is the Equipment Group correct? | A3 | Group I or II + Gas Group |
| Is the Equipment Category correct? | A3 | 1, 2 or 3 |
| Is the Equipment Cable ID as per Loop Diagram? | A3 | |
| Is the IP Rating appropriate to the Group III Material? | A4 | Dust only |
| Is the Equipment Temperature Class correct? | A5 | T1 - T6 |
| Is the Ambient Temperature Range correct for the Installation? | A6 | |
| Is the Service Temperature Range correct for the Installation? | A7 | |
| Is the Installation correctly and clearly labelled? | A8 | |
| Are all Stoppers tight? | A10 | |
| Are all Glands complete and tight? | A10 | |
| Is the Maximum Voltage Um of the Barrier correct? | A17 | Barrier Units |
| Is the Maximum Voltage Um of the Barrier not exceeded? | A17 | Barrier Units |
| Are all of the Cover Bolts tight? | E8 | Not in the Standard |

# Inspection Report Checklist

## Protections Exp

| Report Form Reference: | C96 / A4 | Plant Reference: | A4 / Ground Floor |
|---|---|---|---|
| Inspection Grade: | Close | Zone: | Zone 2 |
| Equipment to be Inspected: | Instrument Pressurised Cabinet A4/P23 | | |
| Responsible Electrical Engineer: | I. Staff | Date: | |
| Name of Inspector (Print): | | Signed: | |

| Check: | Code: | Comments: |
|---|---|---|
| Is the Equipment Group correct? | A2 | Group I or II + Gas Group! |
| Is the Equipment temperature Class correct? | A3 | T1 - T6 |
| Is the Equipment ID correct? | E5 | Not in the Standard |
| Is the Earth Bond tight? | E6 | Not in the Standard |
| Are all Glands tight? | E7 | Not in the Standard |
| Are all the Stoppers tight? | E8 | Not in the Standard |

# Report on Inspection:

| Report Form Ref: | C229/A4 | Plant Ref: | A4 / Ground Floor |
|---|---|---|---|
| Inspection Type: | Close | Zone: | Zone 2 |

| Equipment to be Inspected: | Plant Lighting Circuit 4 |
|---|---|

| Electrical Engineer: | I. Staff | Date: | |
|---|---|---|---|
| Name of Inspector (Print) | | Signed: | |

| Location: | Fault: | Code: | Suggested Remedial Action: |
|---|---|---|---|
| | | | |
| | | | |
| | | | |
| | | | |
| | | | |
| | | | |
| | | | |
| | | | |
| | | | |
| | | | |
| | | | |
| | | | |
| | | | |
| | | | |
| | | | |
| | | | |
| | | | |
| | | | |
| | | | |
| | | | |
| | | | |
| | | | |
| | | | |
| | | | |

# Detailed Inspection

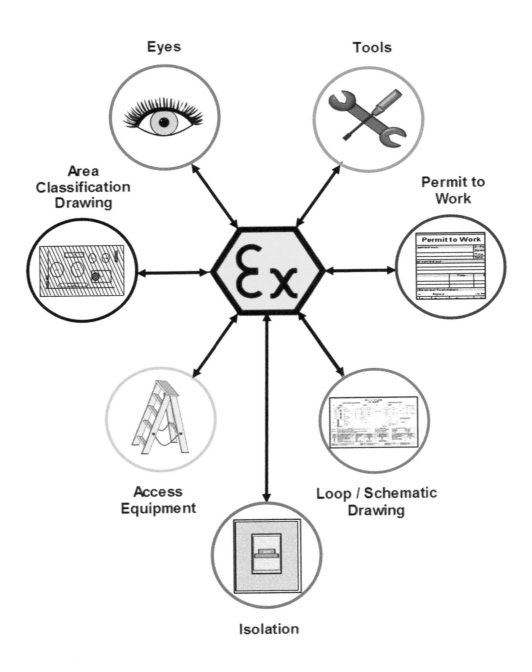

Eyes

Tools

Area Classification Drawing

Permit to Work

Access Equipment

Loop / Schematic Drawing

Isolation

# Detailed Inspection Checklist

The Inspection Checklist lists the points the Inspector looks for when inspecting the equipment e.g. what is the enclosure condition inside. Also with these points are the Codes in the IEC60079 Standard that are applicable to each Inspection Check. On the top of the Inspection Checklist is all of the detail required as to location of the work to be done. Later in the book I have given an explanation and graphic description of **my** interpretation as to the meaning of each of the points on the Checklist and why it must be done. With this document, the Area Classification Drawing, the Method Statement and the Fitting Location Diagram, the Inspector is now ready to inspect the equipment.

The next document to this one in the book is the Report, which the Inspector completes for the engineer regarding the condition of the equipment and any problems that arose from the Inspection. Remember this is a **Detailed** Inspection so an Area Classification Drawing & Schematic/Loop Diagram are available. Access equipment and tools can be used. Other items that can be used are binoculars, thermal image cameras, temperature guns.

FULL ISOLATION PROCEDURE MUST BE CARRIED OUT TO COMPLETE THIS INSPECTION.

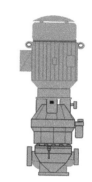

As far as a **Mechanical Detailed** Inspection is concerned I have formulated the Inspection Checklist more on procedures and documentation rather than individual items of equipment. If we take pumps for instance, there are so many, from the standard centrifugal pump up to the Sundyne pumps, example to the right. Just imagine the report form to overhaul this unit, plus they are usually driven by a HV motor so the isolation procedure alone would be quite a task.

So to sum up, companies will have their own report forms for these units which, taking the Sundyne pump (right) as an example, will be quite complex.

The Checklist here centres on whether a Risk Assessment has been completed, what about the isolation procedures, where is the Permit to Work etc. I have also taken into account that much of the mechanical equipment at present is not Atex and in the mechanical world there is no final date for non-Atex equipment to be used.

For Mechanical Isolation, a simple lock off procedure on a Permit to Work will be adequate.

Illustrated on the left is what is called an iso-lock device. There is room on this one for six padlocks, so every person that is working on that particular equipment can put their lock in one of the holes.

The only reason that fuses need to be removed or a full isolation with a circuit breaker carried out is if the electrical connections are being worked upon, e.g., on the removal from site of an electric motor.

# Detailed Inspections
## Documentation Required

In all of the literature on 'Visual' Inspections it says that the only document allowed/required is an Area Classification Drawing. On a 'Detailed' Inspection Schematic/Loop Diagrams are required. Below are several extra documents that may be required on various Inspections depending upon company procedures and policies. Depending upon which Inspection is being carried out, as per title at the top, the documents that may be required on a 'Detailed' Inspection are highlighted below in Red:

1    Permit to Work: This depends solely on company policy, but the chances are that the Inspector(s) would not be allowed on the plant without one.

2    **A Clearance Certificate:** Some companies do not use or rely on these Clearance Certificates and just use a Permit to Work.

3    A Gas Free Certificate: Used when battery & non-IS test equipment is being used or when covers are being opened where there is a potentially live supply. Required on an Electrical Detailed Inspection.

4    A Method Statement: Written by the Electrical, Instrument or Mechanical Engineer stating the safe system of work and will also include the scope of work. (e.g. How far do the Inspectors go?)

5    A Risk Assessment: The Permit will go a long way towards this, but that is a company document whereas this Assessment is completed by the Inspector(s).

6    An Area Classification Drawing: The Inspector(s) would need to know what Zone they are in so they can ensure that the equipment is suitable.

7    An Equipment Location Diagram: The Inspectors(s) would need to know exactly where on the plant the equipment is that they have to inspect.

8    The Checklist: This would show what faults they would be looking for on an Inspection and what codes would go with those faults.

9    The Report: This is the form where the Inspector(s) write down the location, fault, code and suggested remedial action.

10   A Circuit Wiring/Schematic Diagram: If the equipment is a power circuit such as lighting, motors, sockets etc. Usually a Schematic Diagram is used on an Inspection, a Wiring Diagram would not be a great deal of use here and would be more use on an Installation Project.

11   A Loop Diagram: If the equipment is instrumentation the Loop Diagram would be invaluable as all the information on the instruments and barrier units would be on here. This drawing should be as near 100% accurate as possible all of the time.

12   **A Piping and Instrumentation Diagram (P & ID):** This would be required on the Mechanical Inspection to show the pipework included in the Inspection.

# Detailed Inspection

## Isolation Procedure

## Example of Lighting Circuit

**Locate Circuit to be Inspected Junction Box in Zone 2 Area.**

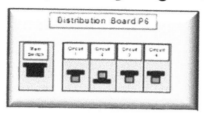

**Locate Distribution Board and Switch off Circuit Breaker.**

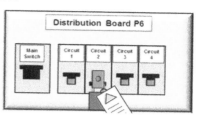

**Fit Lock and Isolation Label on the Circuit Breaker.**

**Fit Label to Switch to let everyone know Circuit is off!**

**Remove the Junction Box Lid to expose the Terminal Block**

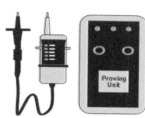

**Test your Potential Indicator on a Proving Unit or Dis-Board**

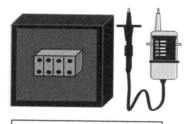

**Test Live to Neutral - Live to Earth - Neutral to Earth! 3 Tests.**

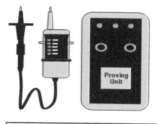

**Re-test your Potential Indicator on a Proving Unit or Dis-Board**

**Temporarily replace the Lid of the JB until you are ready!**

**Permit**
═══════
═══════

**Isolation**

**Sign the Permit in the Isolation Section!**

# Inspection Report Checklist
## Protections Exd, Exe, Exn & Ext

| Report Form Reference: | D111 / A4 | Plant Reference: | A4 / Glound Floor |
|---|---|---|---|
| Inspection Grade: | Detailed | Zone: | Zone 2 |
| Equipment to be Inspected: | | Plant Lighting Circuit 4 from Dis-Board PL2 | |
| Responsible Electrical Engineer: | I. Staff | Date: | |
| Name of Inspector (Print): | | Signed: | |

| Check: | Code: | Comments: |
|---|---|---|
| Is the Equipment Circuit ID correct? | A6 | |
| Is there any Damage or Unauthorised Modifications inside? | A9 | |
| Are Joint faces clean and undamaged? | A13 | Exd Flameproof only |
| Is the condition of the Enclosure Gasket satisfactory? | A14 | |
| Is there any evidence of Water inside of the Enclosure? | A15 | |
| Is there any evidence of Dust inside the Enclosure? | A15 | |
| Are all Electrical Connections tight? | A17 | |
| Are all unused Terminals tightened? | A18 | |
| Are all Enclosed Break devices undamaged? | A19 | ExnC Enclosures only |
| Are all Hermetically Sealed Devices undamaged? | A19 | ExnC Enclosures only |
| Are all Encapsulated Devices undamaged? | A20 | ExnC Enclosures only |
| Are all Flameproof Devices undamaged? | A21 | |
| Are Restricted Breathing Enclosures satisfactory? | A22 | ExnR Enclosures only |
| Are all Test Ports satisfactory? | A23 | ExnR if Fitted |
| Is the Restricted Breathing Operation satisfactory? | A24 | ExnR Enclosures only |
| Is the Lamp Type, Pin Configuration and position correct? | A28 | |
| Is the Electric Motor Winding IR satisfactory? | A31 | Electric Motors only |
| Is the Type of Cable appropriate? | B1 | |
| Are all Stopper Boxes correctly filled? | B4 | |
| Is the integrity of a Conduit in a Mixed System maintained? | B5 | |
| Is all of the Earthing satisfactory inside of the Equipment? | B6 | |
| Is the Fault Loop impedance Reading satisfactory? | B7 | TN-S Systems |
| Is the Earthing Resistance Reading satisfactory? | B7 | IT & TT Systems |
| Are Automatic Protection Devices Set correctly? | B8 | Overloads no auto-reset |
| Do Protection Devices operate within the Set Times? | B9 | Overloads |
| Are Specific Conditions of use complied with? | B10 | X' after Atex Number |
| Are Cables not in use correctly terminated? | B11 | |
| Do Protection Devices operate the tE in permitted time? | B23 | |
| Is Electrical Insulation Clean and Dry? | C3 | |
| Are internal Cables satisfactory? | E5 | Not in the Standard |
| Are the internal Components of Glands correct? | E6 | Not in the Standard |
| Do the Electric Motor Windings Balance? | E7 | Not in the Standard |
| Is the Electric Motor Earth Path Reading satisfactory? | E8 | Not in the Standard |
| Is the Switch Wire marked? | E9 | Not in the Standard |

# Inspection Report Checklist

## Protections Exi

| Report Form Reference: | | D119 / A4 | Plant Reference: | | A4 / 1st Floor |
|---|---|---|---|---|---|
| Inspection Grade: | | Detailed | Zone: | | Zone 2 |
| Equipment to be Inspected: | | Acetic Acid Tank (T23) Level loop: IS / 22 / Lev.DP | | | |
| Responsible Electrical Engineer: | | I. Staff | Date: | | |
| Name of Inspector (Print): | | | Signed: | | |

| Check: | Code: | Comments: |
|---|---|---|
| Are Enclosure Glass / Glass to Metal Seals satisfactory? | A9 | |
| Are there any internal Unauthorised Modifications? | A11 | |
| Is the condition of the Enclosure Gasket satisfactory? | A14 | |
| Are all Electrical Connections tight? | A15 | |
| Are Printed Circuit Boards clean and undamaged? | A16 | |
| Are Cables installed in accordance with Documentation? | B1 | |
| Are Cable Screens Earthed in accordance with Documentation? | B2 | |
| Are Point to Point Connections correct and satisfactory? | B5 | Initial Inspection only |
| Are Earth Connections tight? | B6 | |
| Is Earth Continuity satisfactory? | B6 | |
| Do Earth Connections maintain the integrity of the Installation? | B7 | |
| Is the Intrinsically Safe Earthing satisfactory? | B8 | |
| Is the Insulation Resistance satisfactory? | B9 | Disconnect Equipment |
| Is there at least 50mm between IS and Non-IS connections? | B10 | |
| Is the Short Circuit Protection of the Power Supply suitable? | B11 | |
| Are Cables not in use correctly terminated? | B13 | |
| Any Internal Damage? | E9 | |
| Any Copper showing from Cable Cores/Core End Sleeves? | E10 | |
| Any signs of Water or Dust in the Enclosure? | E11 | |
| Have all Cores got Core End Sleeves fitted? | E12 | Bootlace Crimps |

# Inspection Report Checklist

## Protections Exp

| Report Form Reference: | D121 / A4 | Plant Reference: | A4 / Ground Floor |
|---|---|---|---|
| Inspection Grade: | Detailed | Zone: | Zone 2 |

| Equipment to be Inspected: | Instrument Pressurised Cabinet A4/P23 |
|---|---|

| Responsible Electrical Engineer: | I. Staff | Date: | |
|---|---|---|---|
| Name of Inspector (Print): | | Signed: | |

| Check: | Code: | Comments: |
|---|---|---|
| Is the Equipment Cable/Circuit ID correct? | A4 | |
| Is the Lamp Rating, Type and Position correct? | A9 | |
| Is the Type of Cable appropriate? | B1 | |
| Is the Fault Loop Impedance satisfactory? | B4 | TN-S System |
| Is the Earth Resistance satisfactory? | B4 | IT & TT Systems |
| Do Protection Devices operate within Set Times? | B5 | |
| Are Protection Devices Set correctly? | B6 | |
| Is the Protective Gas Inlet temperature below maximum spec? | B7 | |
| Do Pressure Alarms and Interlocks function correctly? | B11 | |
| Do Flow Alarms and Interlocks function correctly? | B11 | |
| Are Duct Spark and Particle Barriers for exhausting the Gas into a Hazardous Area satisfactory? | B12 | |
| Are Specific Conditions of use complied with? | B13 | 'X' after Atex Number |
| Are Internal Cables satisfactory? | E9 | Not in the Standard |
| Are the Internal Components of Glands correct? | E10 | Not in the Standard |
| Is there any evidence of Water in the Enclosure? | E11 | Not in the Standard |
| Is there any evidence of Dust in the Enclosure? | E12 | Not in the Standard |
| Is the Enclosure Seal satisfactory? | E13 | Not in the Standard |
| Is there any damage to any Components within the Enclosure? | E14 | Not in the Standard |

# Inspection Report Checklist

## Protections Exp Motor

| Report Form Reference: | | D124 / A4 | Plant Reference: | | A4 / Ground Floor |
|---|---|---|---|---|---|
| Inspection Grade: | | Detailed | Zone: | | Zone 2 |
| Equipment to be Inspected: | | Pressurised Motor Tag. Number: P212A | | | |
| Responsible Electrical Engineer: | | I. Staff | Date: | | |
| Name of Inspector (Print): | | | Signed: | | |

| Check: | Code: | Comments: |
|---|---|---|
| Is the Equipment Cable/Circuit ID correct? | A4 | |
| Is the Type of Cable appropriate? | B1 | |
| Do Protection Devices operate within Set Times? | B5 | |
| Are Protection Devices Set correctly? | B6 | |
| Is the Protective Gas Inlet temperature below maximum spec? | B7 | |
| Do Pressure Alarms and Interlocks function correctly? | B11 | |
| Do Flow Alarms and Interlocks function correctly? | B11 | |
| Are Duct Spark and Particle Barriers for exhausting the Gas into a Hazardous Area satisfactory? | B12 | |
| Are Specific Conditions of use complied with? | B13 | 'X' after Atex Number |
| Are Internal Cables satisfactory? | E11 | |
| Are the Internal Components of Glands correct? | E12 | |
| Is there any evidence of Water in the Enclosure? | E13 | Terminal Box |
| Is there any evidence of Dust in the Enclosure? | E14 | Terminal Box |
| Is the Enclosure Seal satisfactory? | E15 | Terminal Box |
| Is there any damage to any Components within the Enclosure? | E16 | Terminal Box |
| Is the Motor Winding IR satisfactory? | E17 | |
| Is the Motor Winding Balance satisfactory? | E18 | |
| Is the Motor Earth Path satisfactory? | E19 | |
| Has the Motor Fan got sufficient clearance? | E20 | |
| Is the Motor Air Flow Clear? | E21 | |

# Report on Inspection:

| Report Form Ref: | D095/A4 | Plant Ref: | A4 / Ground Floor |
|---|---|---|---|
| Inspection Type: | Detailed | Zone: | Zone 2 |

| Equipment to be Inspected: | Plant Lighting Circuit 4 |
|---|---|

| Electrical Engineer: | I. Staff | Date: | |
|---|---|---|---|
| Name of Inspector (Print) | | Signed: | |

| Location: | Fault: | Code: | Suggested Remedial Action: |
|---|---|---|---|
| | | | |
| | | | |
| | | | |
| | | | |
| | | | |
| | | | |
| | | | |
| | | | |
| | | | |
| | | | |
| | | | |
| | | | |
| | | | |
| | | | |
| | | | |
| | | | |
| | | | |
| | | | |
| | | | |
| | | | |
| | | | |
| | | | |

# Detailed Inspections (Mechanical)
## Documentation Required

In all of the literature on **'Visual'** Inspections it says that the only document allowed/required is an Area Classification Drawing. On other Inspections Schematic/Loop Diagrams are required. Below are several extra documents that may be required on various Inspections depending upon company procedures and policies. Depending upon which Inspection is being carried out, as per title at the top, the documents that may be required are highlighted below in Red:

1   Permit to Work: This depends solely on company policy, but the chances are that the Inspector(s) would not be allowed on the plant without one.

2   A Clearance Certificate: Some companies do not use or rely on these Clearance Certificates and just use a Permit to Work.

3   **A Gas Free Certificate:** Not used on Mechanical Inspections, Electrical Visual Inspections and Electrical Close Inspections because no test equipment is being used and no covers opened where there is a potentially live supply. Required on an Electrical Detailed Inspection.

4   A Method Statement: Written by the Electrical Engineer stating the Safe System of Work which will also include the scope of work. (e.g. How far do the Inspectors go?)

5   A Risk Assessment: The Permit will go a long way towards this, but that is a company document whereas this Assessment is completed by the Inspector(s).

6   An Area Classification Drawing: The Inspector(s) would need to know what Zone they are in so they can ensure that the equipment is suitable.

7   An Equipment Location Diagram: The Inspectors(s) would need to know exactly where on the plant the equipment is that they have to inspect.

8   The Checklist: This would show what faults they would be looking for on an Inspection and what codes would go with those faults.

9   The Report: This is the form where the Inspector(s) write down the location, fault, code and suggested remedial action.

10  **A Circuit Wiring/Schematic Diagram:** If the equipment is a power circuit such as lighting, motors, sockets etc. Usually a Schematic Diagram is used on an Inspection, a Wiring Diagram would not be a great deal of use here and would be more use on an Installation Project.

11  **A Loop Diagram:** If the equipment is instrumentation, the Loop Diagram would be invaluable as all the information on the Instruments and Barrier Units would be on here. This drawing should be as near 100% accurate as possible all of the time.

12  A Piping and Instrumentation Diagram (P & ID): This would be required on the Mechanical Inspection to show the Pipework included in the Inspection.

## Inspection Report Checklist

## Mechanical Inspection

| Report Form Reference: | DM225/A4 | Section: | A4 Mechanical |
|---|---|---|---|

| Procedure for carrying out a Detailed Inspection of Mechanical equipment: | | | |
|---|---|---|---|

| Mechanical Engineer: | H. Cole | Date: | |
|---|---|---|---|
| Name of Inspector (Print): | | Signed: | |

| Check: | Comments: | Tick or N/A |
|---|---|---|
| Has the equipment been identified as the correct unit? | Engineering | |
| Has it been arranged with process to carry out the work? | Eng. / Process | |
| Is it agreed with the workshop to do the work? | Workshop | |
| Is the equipment involved in the overhaul Atex? | Atex | |
| If above answer is yes, is the equipment Exh, Exd or Exp? | Atex | |
| Have Technicians involved got an Atex Certificate? | Atex | |
| Has a Method Statement (Safe System of Work) been done? | Method State. | |
| Have all spares been ordered ready for the overhaul? | Spares | |
| Has a crane been organised to lift the equipment out? | Crane | |
| Has a trailer been organised to transport the equipment? | Transport | |
| Above vehicles may be subject to a Gas Free Certificate? | Gas Free Cert. | |
| Has a Risk Assessment been completed by Team Leader? | Risk Assessment | |
| Has a Permit to Work been Issued/signed to do the work? | Permit T. W. | |
| Have all involved in the work signed on the Permit to Work? | Permit T. W. | |
| Has a Clearance Certificate been issued for isolations? | 6 Items Below: | |
| Have all electric drives involved been isolated? | Clearance Cert. | |
| Have all associated valves been locked closed? | Clearance Cert. | |
| Has equipment involved been drained of hazardous liquid? | Clearance Cert. | |
| Has any pipework involved been drained of hazardous liquid? | Clearance Cert. | |
| Has the equipment/pipework been steamed out? | Clearance Cert. | |
| Technicians/Supervision must sign the above certificate. | Clearance Cert. | |
| If Clearance Certificate is non tearable attach to the equipment. | Company Policy? | |
| Has an estimated time been agreed for the overhaul? | Time | |
| If there are standby units are they working satisfactorily? | Standby Unit | |

I have formulated the above Checklist for the Mechanical Detailed Inspection more as a procedural exercise than a Detailed Inspection of any particular piece of equipment such as a pump etc., as these overhauls will be standard specialised units and companies will formulate their own checklists for these, possibly kept on a Company Database.

# Report on Mechanical Inspection:

| Report Form Ref: | DM207/A4 | Plant Ref: | A4 / Ground Floor |
|---|---|---|---|
| Inspection Type: | Detailed | Zone: | Zone 2 |

| Equipment to be Inspected: | Column C101 feed Pump P10A |
|---|---|

| Mechanical Engineer | I. Staff | Date: | |
|---|---|---|---|
| Name of Inspector (Print) | | Signed: | |

| Location: | Fault: | Code: | Suggested Remedial Action: |
|---|---|---|---|
| | | N/A | |
| | | N/A | |
| | | N/A | |
| | | N/A | |
| | | N/A | |
| | | N/A | |
| | | N/A | |
| | | N/A | |
| | | N/A | |
| | | N/A | |
| | | N/A | |
| | | N/A | |
| | | N/A | |
| | | N/A | |
| | | N/A | |
| | | N/A | |
| | | N/A | |
| | | N/A | |
| | | N/A | |
| | | N/A | |
| | | N/A | |
| | | N/A | |

# Power Code Descriptions

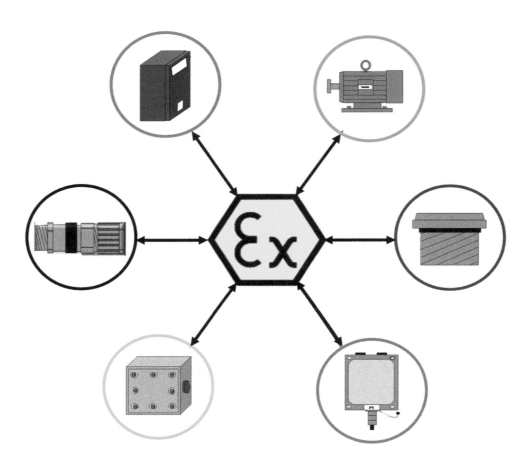

# Visual Inspection
## Is the Equipment Appropriate to the EPL/Zone Requirements of the Location?

## Code A1

### Introduction:

All that the Inspectors will have with them here is an **Area Classification Drawing.** This is a drawing with lines, circles or squares on it depicting the Zones. **No tools or access equipment.**

### Area Classification Drawing Markings:

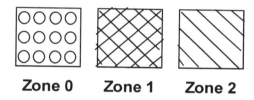

**Zone 0    Zone 1    Zone 2**

The **Gas Zone** markings will be shown on the **Area Classification Drawing** as shown to the left. These days Zone 2 will be the most common. Zone 0 usually inside of tops of vessels.

The **Dust Zone** markings will be shown on the **Area Classification Drawing** as shown to the right. Zone 20 usually inside tops of hoppers etc. Because of PPM Zone 21 areas will be rare.

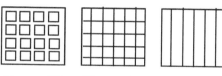

**Zone 20    Zone 21    Zone 22**

On this drawing there will be no Gas or Dust Groups, no Temperature Classifications or Surface Temperatures so although this information will be on the equipment there will be no reference on the drawing to match up.

### Protection/Equipment Protection Level (EPL):

What the Inspectors are looking for here is the 'Protection' i.e. Exd, Exe, Ext, etc. on the equipment and to see if it is in the correct Zone **(see chart on next page).**

Manufacturers have started putting the EPL on the end of the markings probably to get technicians and Inspectors used to seeing them. These are, of course, being introduced by the IEC probably with a view to replacing the category. Gas Zone EPLs are Ga, Gb or Gc. As can be seen from the above diagram this **EPL is 'Gb'** i.e. suitable for **Gas Zone 1.** Dust EPLs will be Da, Db or Dc.

It must be noted that equipment suitable for Dust Zones will not automatically be suitable for Gas Zones and vice-versa. So if, say, the Inspector was to find **EPL Gb** in a **Dust Zone 21** they should mark this as a fault.

# Visual Inspection

## Is the Equipment Appropriate to the EPL/Zone Requirements of the Location?

### Code A1-1

| Protection: | Description: | Special Comments: | Zones: |
|---|---|---|---|
| Exd | Flameproof | Allows gas to enter | 1 & 2 |
| Exe | Increased safety | IP54 Minimum | 1 & 2 |
| Exec | Increased safety | Replaced **ExnA** | 2 ONLY! |
| Exh | Mechanical | Combines Mechanical 'c' - 'b' - 'k' | Documentation |
| Exia | Intrinsic Safety | 2 Faults and remain safe. | 0, 1 & 2 |
| Exib | Intrinsic Safety | 1 Fault and remain safe. | 1 & 2 |
| Exic | Intrinsic Safety | Replaced **ExnL** | 2 ONLY! |
| ExiD | Intrinsic Safety | Dust Atmospheres | Documentation |
| Exm | Encapsulated | Older type equipment | 1 & 2 |
| Exma | Encapsulated | Completely seals | 0, 1 & 2 |
| Exmb | Encapsulated | Completely seals | 1 & 2 |
| Exmc | Encapsulated | Completely seals | 2 ONLY! |
| ExmD | Encapsulation | Dust Atmospheres | Documentation |
| ExN | Non Incendive | Never Cenelec so no export! | Withdrawn |
| ExnA | Reduced Risk | Non Sparking - Replaced by **Exec** | 2 ONLY! |
| ExnC | Reduced Risk | Encapsulated | 2 ONLY! |
| ExnC | Reduced Risk | Hermetically Sealed | 2 ONLY! |
| ExnC | Reduced Risk | Enclosed Break | 2 ONLY! |
| ExnR | Reduced Risk | Restricted Breathing | 2 ONLY! |
| ExnL | Reduced Risk | Energy Limiting - Replaced by **Exic** | 2 ONLY! |
| ExnZ | Reduced Risk | Pressurisation - Replaced by **Expz** | 2 ONLY! |
| Exo | Oil Filled | Quenches any arcs or sparks | 1 & 2 |
| Exop (is) | Optical Radiation | Inherent Safety (is) | 0, 1 & 2 |
| Exop (sh) | Optical Radiation | Not Inherently Safe and Interlocked (sh) | Documentation |
| Exop (pr) | Optical Radiation | Protected Optical System (pr) | Documentation |
| Exq | Quartz / Powder Filled | Quenches any arcs or sparks | 1 & 2 |
| Expx | Pressurisation | Auto Shutdown | 1 & 2 |
| Expy | Pressurisation | No Shutdown | 1 & 2 |
| Expz | Pressurisation | No Shutdown - Replaced **ExnZ** | 2 ONLY! |
| Expv | Pressurisation | Auto Shutdown | Documentation |
| ExpD | Pressurisation | Dust Atmospheres | 21 & 22 |
| Exs | Special Protection | Older type equipment! Never Cenelec | Withdrawn |
| Exsa | Special Protection | New Atex Certification | 0, 1 & 2 |
| Exsb | Special Protection | New Atex Certification | 1 & 2 |
| Exsc | Special Protection | New Atex Certification | 2 ONLY! |
| ExtD | Protection by Enclosure | Dust Atmospheres only! | Documentation |
| Exv | Ventilation | American Ex Standard | Documentation |

# Close Inspection
## Is the Equipment Group Correct?

Code A2

Introduction:

What the inspector is looking for here is the **'Group'**. This can take several forms e.g. Surface Industry or Mining, the **Gas Group** and the **Dust Group**. If the equipment is Atex the Group will be in the form of Roman numerals after the Atex mark 'l' or 'll'. The predominant gases in each Gas Group are llA-Propane, llB-Ethylene & llC-Hydrogen. Dust Group equipment cannot, without the manufacturer's approval in writing, go into Gas Zones and vice versa.

Mining Equipment:

If the equipment is for mining, the **'Group'** is the **Roman l** after the Atex mark as shown in the diagram to the left. 'Ma' indicates the top section in mining.

Surface Industry Gas/Vapour Equipment:

As you can see in the diagram to the right the **'Group'** is the **Roman ll** surface industry, after the Atex mark.

Gas Group ll:

The diagram on the left shows the **Gas Group** as just a **Roman ll.** Not to be confused with the **ll** for surface industry.

If the **Gas Group** is shown as just a **Roman ll** then the equipment will be suitable for any **Gas Group ll.** If the equipment is shown as **Gas Group llB + H2** then it is a **llB** piece of equipment, but also suitable for **Hydrogen** (Not Acetylene or carbon di-sulphide.) Find out the Gas Groups of all chemicals, gases & vapours on plant and ensure equipment Gas Group is correct, (which should be to the worst one) **llC being the worst.**

Surface Industry Dust Equipment:

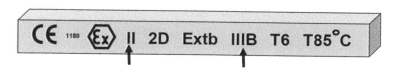

The **Group** will still be a **Roman ll** surface industry even though the Dust Groups are 'lll'. There is Dust Group **lllA** (Combustible flyings e.g. jute, hemp, oakum, kapok etc) **lllB** (Non conductive dust e.g. wood, grain, starch, sugar etc), or **lllC** (Conductive dust e.g. coal, carbon, charcoal, magnesium etc.) **lllC being the worst.**

# Close Inspection
## Is the Equipment Temperature Classification Correct?

Code A3

Introduction:

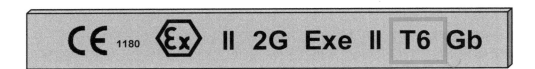

This of course cannot be completed until the **Close** Inspection as the information required would not be on the **Area Classification Drawing.** In this country we have six **Temperature Classifications.** The definition is: the surface of the equipment in contact with the gas, vapour or dust will not go over the **Temperature Classification** (T1 – T6) **under normal or specified fault conditions.** (It does not say it will never go over and the manufacturers will specify the faults.)

Ignition Temperature:

Find out the **Ignition Temperatures** of all of the chemicals, gases, vapours and dusts on the plant and ensure that the **Temperature Classifications** of all of the items of electrical equipment are below that value. In reality Electrical Engineers will order T5 or T6 equipment to be on the safe side even though this may be more expensive.

| | |
|----|-----|
| T1 | 450 |
| T2 | 300 |
| T3 | 200 |
| T4 | 135 |
| T5 | 100 |
| T6 | 85 |

For example if you had a gas or vapour that had an **Ignition Temperature** of 235C. Looking at the UK chart, left, it would be no good using equipment with **Temperature Classification** T2 (300C) because if the equipment was to reach 300C under normal or specified fault conditions and there was a gas cloud with a 235C **Ignition Temperature** the 300C equipment would ignite it.

The **Temperature Classification** of the equipment must be **under** the **Ignition Temperature** of the gases in which it is installed.

Ambient Temperature:

The above definition applies in a normal **Ambient Temperature of -20C to +40C.** So if the equipment does not specify anything else on the data plate then this is the **'Norm'.**

But what if the Ambient Temperature where the equipment is to be installed is higher or lower than the **'Norm'**? This situation needs different equipment so it's back to the Manufacturers to obtain this new equipment. The data plate will state the higher or lower **Ambient Temperature** requested with **'Ta'** or **'Tamb'** showing what the new higher or lower Ambient Temperature is, e.g. **Ta = -50C to +55C.**

# Close Inspection
## Maximum Surface Temperature?
## (Dust Equipment Only.)

Introduction:

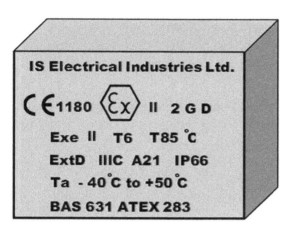

With equipment which is designed for the dust world there is a physical temperature given on the equipment besides T1 – T6 and that is the maximum surface temperature based on a **5mm** layer of dust.

Layers beyond 5mm must be reported and the equipment layer specification checked very carefully.

Dust Ignition Temperatures:

If we just work on a golden rule that 'clouds explode and layers burn', the table (right) shows the **Ignition Temperatures** of various dusts, both their clouds and layers.

Our equipment temperature must not get anywhere near to these temperatures or there could be an explosion (Cloud) or fire (Layer).

The formula is that the surface temperature must be **75C** below a dust **Layer Ignition Temperature.** Remember that these parameters are based on a **5mm layer of dust.**

The surface temperature must be **2/3rds (66%)** of a **Cloud Ignition Temperature.** Excessive dust or dirt beyond the 5mm could smother the equipment and cause it to heat so good housekeeping is vital. **As mentioned above, poor housekeeping should be reported as a fire risk.**

| | Ignition Temperatures °C | |
| --- | --- | --- |
| Dust | Cloud | Layer |
| Cellulose | 520 | 410 |
| Flour | 510 | 300 |
| Grain | 510 | 300 |
| Sugar Dust | 490 | 460 |
| Tea Dust | 490 | 340 |
| Starch | 460 | 435 |
| Lignite | 390 | 225 |

# Visual Inspection
## Is the Degree of IP Satisfactory?

Code A5

### Introduction:

**Ingress Protection** of the equipment is extremely important as this controls whether or not water, gas, vapour or dust can penetrate the equipment. Usually electrical equipment is **IP54** minimum. The IP relies on the quality of the seals on the lids and gland plates of the equipment.

### IP Code:

| Solids | | Liquids | |
|---|---|---|---|
| 0 | No Protection | 0 | No Protection |
| 1 | Hands | 1 | Drip Proof |
| 2 | Fingers | 2 | Shower Proof |
| 3 | Tools | 3 | Rain Proof |
| 4 | Wires | 4 | Splash Proof |
| 5 | Dust | 5 | Hose Proof |
| 6 | Fine Dust | 6 | Ships Deck |
| | | 7 | Under Water |
| | | 8 | Under Water |

The first digit refers to solids and the second digit to liquids. E.g. IP54 (the liquid of course being water.) I have indicated IP54 on the diagram to the left.

The solids range from fine dust to 'no protection' and the liquids range from totally under water to 'no protection'.

On the equipment to the right the IP is IP66 as many JBs are these days. Manufacturers no longer have an obligation to display the IP Rating on their equipment.

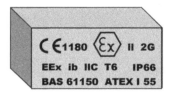

### IP and Cable Glands:

There may be an IP rating on a cable gland, that refers to the level of sealing effectiveness of the outer seal of the gland where it is on the sheath of the cable as stated in IEC60529, and has nothing to do with the IP rating of the equipment where the gland is to be installed.

### Fibre & Nylon IP Washers:

I am often asked when I am CompEx training why sometimes nylon and sometimes fibre washers are used. Care must be taken with fibre washers at lower temperatures as they tend to absorb water and at lower temperatures, they can then freeze and crack. Also fibre washers have a finite life.

Nylon IP washers, sometimes known as 'Entry Thread Washers' have an operating temperature of around -60C to +150C so can go down to very low temperatures. They can distort after use and may not be as effective if re-used.

# Visual Inspection
## Is the Degree of IP Satisfactory?

## Code A5-1

### IP54, IP55 & 6mm Thread Length:

Inspectors will not, obviously, be able to see seals on doors and lids until they come to do the **Detailed** Inspection, but they may be able to see if gland accessories are in place such as IP washers. In an **IP54 Area** on Exe, Exn and Exi equipment the gland thread of the box has to be under **6mm** long before an IP washer is required. **So 6mm or over, no IP Washer.** In an **IP55 Area** on Exe, Exn and Exi equipment, IP washers are required regardless. **See Section A5-1.**

### IP washers on Equipment with Threaded Holes:

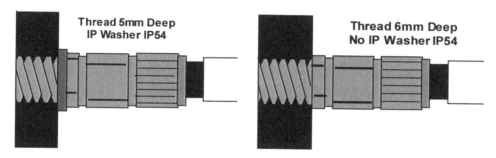

Thread 5mm Deep
IP Washer IP54

Thread 6mm Deep
No IP Washer IP54

Let us say Inspectors are inspecting an Exe junction box, they have no idea of the length of thread on the Gland entries until the **Detailed** Inspection. They are not interested in the thickness of the box although sometimes the thread length & thickness are the same. They could look at the other entries to see if they have IP washers fitted. If in doubt, my advice is to mark down as missing.

**THIS DOES NOT APPLY TO RESTRICTED BREATHING EQUIPMENT.**

### Clearance Holes:

Clearance holes require IP washers. If you do not put them on here then water will just run down the gland thread and into the equipment. On the right is an example of a clearance hole with an IP washer on the outside and a serrated washer and locknut on the inside. If an earth tag is fitted the IP washer is fitted next to the equipment.

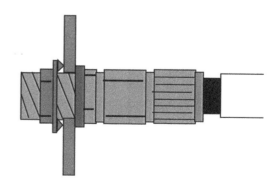

# Visual Inspection
## Is the Degree of IP Satisfactory?

## Code A5-2

### IP Washers and Flameproof Equipment:

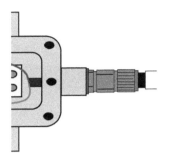

Exd Flameproof equipment should not require IP washers. There is no legislation stating that they cannot be fitted if it is company policy. Companies might contact the manufacturers to see what their views are on this practice.

Inspectors must ensure that where glands are fitted, **FIVE** full threads of the gland are entered into the equipment.

### IP Washers and Restricted Breathing:

If Inspectors are checking a light fitting which is **ExnR Restricted Breathing** they must ensure that the manufacturer's IP washers have been used on the gland. The washers are making an IP54 minimum seal.

Victor Fittings insist that on their restricted breathing bulkhead fittings their special IP washers are used, which consist of a thinner rubber washer next to the fitting and a larger metal washer to prevent the gland damaging the rubber as it is tightened.

**PLEASE REMEMBER THIS IP WASHER ARRANGEMENT ONLY APPLIES TO VICTOR LIGHTING. OTHER MANUFACTURERS WILL INDICATE WHICH METHOD IS THEIRS.**

### IP with Stoppers

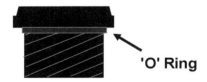

**'O' Ring**

Stoppers used in ExnR light fittings must be the type specified by the manufacturers or ones that came with the fitting. Usually the ones with 'O' rings are fitted. (Exe stoppers will usually have this 'O' Ring.)

### IP Washers above IP54:

If the **area** is more than IP54 e.g. IP55, and the junction box has an adaptor/reducer before the gland, the adaptor/reducer must have an IP washer as it enters the box and the gland must have an IP washer as it fits into the adaptor/reducer.

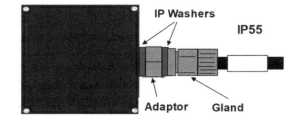

# Visual Inspection
## Is the Degree of IP Satisfactory?

### Code A5-3

#### Higher & Lower IP:

There have been cases in the past where the client has stated that equipment be, say **IP66,** but the installation company has installed equipment to **IP68** thinking it was a better IP and would therefore be acceptable, and has then had to replace it with equipment to **IP66 as the client requested.** Equipment to **IP68**, apparently, is tested in a different way to **IP66**.

#### IP Washers with Exp Pressurisation:

We will be discussing this further in the Exp pressurisation section, but on Exp equipment the gland, in many cases, does not need to be Certified, but must fit the cable correctly and seal into the equipment. Usually with pressurisation we are more concerned with the outside atmosphere leaking in than pressurisation gas leaking out. So at least the pressurisation gas will be at positive pressure.

#### IP with Dusts:

If the equipment is just for dust atmospheres it is likely that there is no second number and it has been replaced by an 'X' as in the diagram left. So instead of IP54 or IP66 it would be **IP5X or IP6X.** What they are saying here is that if the equipment is suitable for dust it is automatically suitable for liquids.

**THIS APPLIES ONLY IN THE DUST WORLD AND DOES NOT APPLY TO GASES AND VAPOURS.**

#### Dusts: The Three Zone System:

Based on the three Zone approach, IP washers would be required on dust area equipment as shown on the right. We can take the third Zone one step further and say if the Dust is **CONDUCTIVE,** IP washers are required, but not if the Dust is **NON-CONDUCTIVE.**

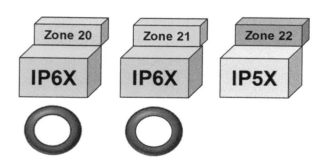

# Visual Inspection
## Is the Degree of IP Satisfactory?

### Could there be a Third Number:

| IK Number: | Impact Energy in Joules: | Equivalent to: |
|---|---|---|
| 0 | Unprotected | 200g dropped from 7cm |
| 1 | 0.15 | 200g dropped from 10cm |
| 2 | 0.2 | 200g dropped from 17.5cm |
| 3 | 0.35 | 200g dropped from 25cm |
| 4 | 0.5 | 200g dropped from 35cm |
| 5 | 0.7 | 500g dropped from 10cm |
| 6 | 1 | 500g dropped from 20cm |
| 7 | 2 | 500gm dropped from 40cm |
| 8 | 5 | 1,7kKg dropped from 29.5cm |
| 9 | 10 | 5Kg dropped from 20cm |
| 10 | 20 | 5Kg dropped from 40cm |

The answer is yes. At one time there was definitely going to be a third number and that was going to be the amount of Impact Energy in Joules that the equipment could take.

The Euro Norm for this is EN62262 which was the 'IK' Code with the equivalent International Standard being IEC62262 (2002).

The table left shows the detail. In my experience I have not seen many items of equipment with a third number.

### Other Letters that may appear on Equipment:

| IP Code: | Equipment Protected from: |
|---|---|
| IP6K | Powerful jets of Water with Increased Pressure. |
| IP9K | Powerful High Temperature Water Jets |

The letters to the left may appear if High Pressure jetting equipment is to be used as this would go beyond the normal pressure that the item of equipment could stand in a normal test situation.

Other miscellaneous letters which may appear on the IP Protection Code are shown on the right. These, to my knowledge are not common, but could be there.

| Letter: | Protection: |
|---|---|
| C | Protection against Tool access |
| D | Protection against Wire access |
| f | Oil Resistant |
| H | High Voltage |
| M | Equipment Moving in Water Test |
| S | Equipment Still in Water Test |
| W | Weather Conditions |

# Detailed Inspection
## Is the Equipment Circuit ID CORRECT?

## Code A6

### Introduction:

Circuit numbers are vital where isolations are concerned. Now on the **Visual** Inspection Inspectors could only see if the numbers were **present**, but on a **Detailed** inspection they can see if they are **correct** because of extra documentation. Some might argue that this is a **Close** Inspection, but possibly the standard thinking is that with a **Detailed** Inspection the numbers may have to be used in anger to isolate so they can be checked at the same time?

### How to decide what is required:

What do companies actually include in their numbers? In the diagram on the left the cable number is **P4 C3** meaning that this equipment is fed from **distribution board P4 and circuit 3**. The number can be made too long of course, and then it may be confusing.

If the plant is new, there can be situations where the cable numbers are incorrect as they may not have been used in anger for isolations, but if it is a well-established plant the chances are that the numbers have been used before so there is more chance of them being correct. Obviously the cable must have a clear circuit number on either end.

### Is the Box Number and Circuit Numbers Present?

Marshalling boxes are large junction boxes out on the plant. There can be hundreds of different boxes with a whole range of different circuits going to the plant instrumentation e.g. IS Boxes, 24V DC Boxes etc. How these boxes are numbered is for plant/platform policy to decide.

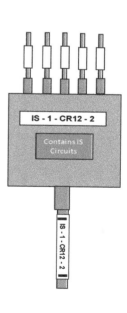

The number might include the box's own number which in the diagram on the right would be IS-1, (intrinsically Safe Circuits hence it must have a blue label on stating that the box only contains IS circuits/loops), and where the main cable comes from/goes to. In the diagram on the right this is Control Room 12 (CR12), the cabinet number where the main cable comes from/goes to in the diagram is number 2. So the number is: **IS – 1 – CR12 – 2**. All of the loop cables must also have individual numbers.

On a **Close/Detailed** Inspection Inspectors can see if the numbers are correct by the extra documentation that they now have.

# Visual Inspection
## Is the Equipment Circuit ID AVAILABLE?

## Code A7

### Introduction:

Circuit numbers are vital where isolations are concerned. At this moment all the Inspector is interested in on a **Visual** Inspection is that the number is there, **not** if it is correct. Remember he/she would only have an Area Classification Diagram.

### What to look for:

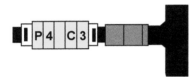

The diagram to the left shows a clear number which should of course be on either end of the cable. There is no documentation available at this point on a **Visual** Inspection stating if the number is correct.

Numbers fall off or become damaged and unreadable, in this case this is a fault and should be recorded. When there are cable numbers and isolations are performed, technicians would be 95% sure that they have the correct circuit so that when they open up boxes on site to, say, check isolation, they are reasonably sure that the correct circuit is isolated. The other 5% is proven with a potential indicator. With no circuit number this percentage will drop dramatically and it could end up with the wrong box being opened.

### Are the Box Number and Circuit/Loop Numbers Present?

Marshalling boxes are large junction boxes out on the plant. There can be hundreds of different boxes with a whole range of different circuits going to the plant instrumentation e.g. IS Boxes, 24V DC Boxes etc.

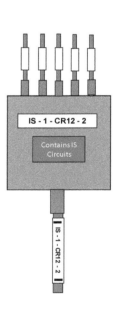

In the case of a similar box to the one on the right, does the main incoming cable have a number? Do each of the individual Loop cables have numbers?

Whilst the Inspector is there, does the box itself have a number and, if it is an IS box, is there a blue label present informing people that it contains intrinsically safe circuits. Again if this box is 'IS' only, are all of the cables blue? If not, are they marked? This is also covered in another section.

To sum up, all the Inspectors are interested in on a **Visual** Inspection is: are all numbers **present**? They have no **Loop Diagram** to check if they are right.

# Visual Inspection
## Is there any Damage to Glass to Metal Seals?

## Code A8

### Introduction:

Glass to metal seal damage is very important as gas, vapour, dust or even water could impregnate the fitting through the damage. This seal should never be bodged up. Always if possible replace with a manufacturer's seal or buy a new unit.

### Has the fitting got a damaged Metal – Glass Seal?

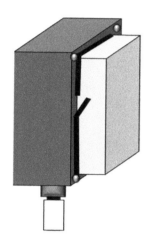

If the glass to metal seal is visible as in the diagram left, check if the fitting is dangerous, be this rubber or compound seal.

There have been instances where the seal has been damaged and water has entered the enclosure and then someone has drilled a hole in the wall of the enclosure to let the water out.

In some cases Inspectors will not discover any damage until they come to the **Detailed** Inspection, but on the example to the left the seal is clearly damaged.

If the fitting is restricted breathing this will be a problem.

### Bodging up Metal to Glass Compound:

Sometimes, especially on older flameproof floodlights, a hardened compound is used to seal the glass to the metal enclosure. Over the years some may crack and fall out.

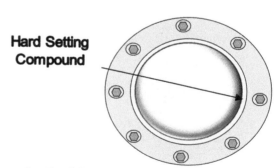

Hard Setting Compound

So the Inspection is to check that there has been no patching up of the sealing compound with an unknown hard setting compound.

Remember, some of these floodlights we are talking about may be Exd flameproof.

The compound used to hold in the glass by the **MANUFACTURER** has been tested by a **Notified Body** such as **BASEEFA** as part of the equipment. Any compound that is bodged up **WILL NOT HAVE BEEN TESTED AS PART OF THE FITTING.**

# Detailed Inspection
## Is there any Damage or Unauthorised Modifications Inside of the equipment?

## Code A9

### Introduction:

Electrical isolation required and Gas Free Certificate. On a **Visual** Inspection the outside of the equipment is all that can be checked for damage, but now that the equipment has been opened up on the **Detailed** Inspection, flange faces, screw threads & gaskets can be checked. I have selected a junction box for this section.

### Are all Exd Threads Clean & Undamaged?

Ensure that all of the flameproof threads are clean and free from damage. **DO NOT FORGET THE EQUIPMENT SCREW LID. Even the slightest mark should be reported.** On closing the enclosure ensure that a smear of non-setting grease is applied to the threads both on the equipment and on the lid. The lid should be screwed back on with the hexagonal grub screw in the right position which is usually 6 o'clock or 12 o'clock and tightened **'Spanner Tight'.**

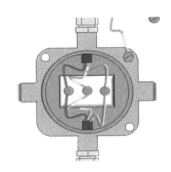

### Is the Terminal Block Undamaged and the correct Manufacturer's?

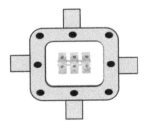

Ensure that the terminal block is the manufacturer's and the correct one for the equipment and not an unauthorised strip terminal block as per the box on the left. The terminal block should be fastened into the equipment as per manufacturer's instructions. Is the terminal block itself damaged at all? What tends to be the case is that too large a screwdriver is used to fit into the terminals and they crack or chip.

### Is the Gasket in good condition?

Sometimes flameproof equipment has a gasket or an O ring, which must not encroach upon the flameproof face of the equipment should it get worn or fragmented which would, of course, make the flameproof gap too large. This gasket should be a manufacturer's gasket and not one that someone has cut out of a sheet of rubber. Another most important point is that a gasket or O Ring must never be left off if the equipment is meant to have one.

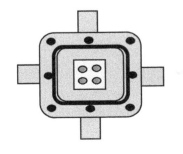

# Visual Inspection
## Is there any Evidence of Unauthorised Modifications Outside of the Equipment?

## Code A10

### Introduction:

One of the problems can be illustrated if we take an example of a Certified ExnR bulkhead light. The manufacturers state that this fitting should be installed in a particular way. They stipulate what glands, stoppers and IP protection to use. If there was an explosion and it was found that this light fitting was the cause and the technician had installed the fitting **EXACTLY** to their specification, the explosion would be the **MANUFACTURER'S** fault. If the fitting had not been installed **EXACTLY** as per specification and the technician had carried out an unauthorised modification, which might be using different IP washers, or the wrong stoppers etc., in court the manufacturer could put some or all of the blame on the way it was installed so it would be the **TECHNICIAN'S** fault. So **NEVER** modify Certified equipment. If you do, the Certificate of Conformity becomes null & void. People do things with the best intentions, but on the face of it some are quite dangerous. Let us look at several examples below. I am sure you can think of more.

### Example: Drilling Holes in Certified equipment:

Drilling holes in the equipment. As I have mentioned, sometimes people do things with the best intentions that end in a disaster. If you look at the picture, left, you will see I have marked where a hole has been drilled near the bottom. Someone may have found water inside in the past due to a faulty seal and treated the symptom, but not the cause, in drilling a hole to let the water out should it get in again. This action of course is dangerous. **There is an obvious seal problem that should be rectified.**

### Example: Drilling holes in a cover to fasten a label:

Self Tapping screws

Drilling the cover to fasten a label on. This of course must be discouraged. Take the Exe junction box lid on the left, the Atex label will be attached by the manufacturers with specially designed fixing units.

Some companies will put on their designate number e.g. JB1 and sometimes drill the lid and attach the label with self-tapping screws.

# Visual Inspection

## Is there any Evidence of Unauthorised Modifications Outside of the Equipment?

## Code A10-1

### Example: Uncertified Stoppers etc. into Certified Equipment:

Fitting uncertified items into Certified equipment such as glands, stoppers etc is entirely against the equipment specification, or any manufacturer's instructions and should be discouraged. These stoppers may not offer the IP quality or protection to which the box is Certified. Always ensure that Certified glands etc. are fitted to Certified equipment.

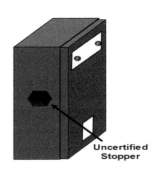

**Uncertified Stopper**

### Example: Exd to Exd or Uncertified with Conduit & no Stopper Box:

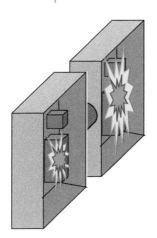

Causing pressure piling with Exd flameproof equipment by joining two pieces of Exd flameproof equipment, as shown left, with a conduit and no stopper box. An explosion in the box on the left will travel through the conduit compressing the gas in front of it and the explosion in the box on the right will be much larger.

Causing pressure piling by joining Exd flameproof equipment to, say, Exe increased safety equipment, as shown left, by conduit with no stopper box. An explosion in the Exd flameproof box on the left will travel down the conduit and blow the box on the right to pieces as Exe equipment is not made to withstand explosions from within.

### Example: Fitting Plastic Exe Stoppers in Exd Equipment:

Exe increased safety plastic stoppers must not be fitted in Exd Flameproof equipment. Although these stoppers are Atex Certified, they will just blow out should there be an explosion inside the equipment and allow the explosion heat and pressure out. Exe stoppers offer mainly IP protection only.

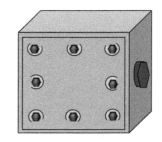

# Visual Inspection

## Is there any Evidence of Unauthorised Modifications Outside of the Equipment?

### Code A10-2

Example: Clips broken on Fluorescent:

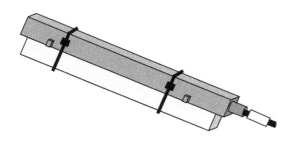

The holding clips have become brittle and broken on an Exe fluorescent lamp diffuser, left, so the technician has put some cable ties around to hold it in place.

Remember the Certificate of Conformity - we must not modify in any way.

The clips are part of the equipment the cable ties are not. Using cable ties, or any incorrect fastening devices in this way may seem effective but de-certify the fitting.

Example: Put Tape over a crack in the di-fuser:

Putting tape over a crack in the di-fuser is an unauthorised modification. How can equipment guarantee IP54 if actions such as this are carried out?

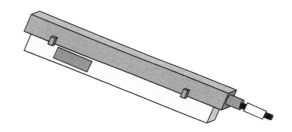

Example: Using Lid Bolts to fasten Clips:

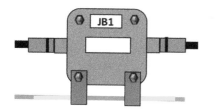

What can be classed as an unauthorised modification is one or two of the cover fastening bolts be used to, say, secure cable clips as in the example shown on the left. This would not ensure the **'Spanner Tightness'** of the fixing bolts that is a must.

Example: Not Manufacturer's Bolts:

We discussed this in other sections. When you inspect the lid you find that several bolts/screws are different to the others. Usually the bodge is made with something like roofing bolts which, unfortunately, seem to fit many thread sizes. These bolts are not the original manufacturer's bolts and have not been tested with the equipment.

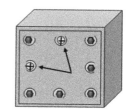

# Visual Inspection

## Is there any Evidence of Unauthorised Modifications Outside of the Equipment?

### Code A10-3

### Example: Not Manufacturer's Grub Screw:

To get into a flameproof piece of equipment technicians or Inspectors would require special tools e.g. sockets, hexagonal keys etc., so screwdriver slotted screws must be wrong. This of course includes the grub screw in a screwed lid. It must be a hexagonal key.

### Example: Wrong size Earth Bond Wire:

Earth Bond wiring has to be a minimum of 4mm². If the bond is only something like 1.5mm² or 2mm² then that is insufficient and Inspectors should list as a fault. If the item is flameproof Exd then there will, in most cases, be a dedicated bonding screw located somewhere on the case. If the item is Exe or Exn then usually this will include an earth tag.

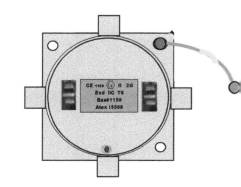

### Different Front:

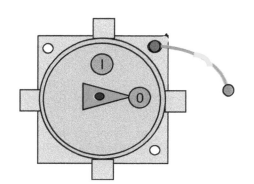

If the enclosure is a flameproof switch as on the diagram on the left, it is possible for the lid on the junction box above to fit. So you might think that, if required, the switch enclosure could be changed simply by removing the switch mechanism, and replacing with a connector block and changing the switch front for a JB lid. This of course would be an unauthorised modification and thus would void any Certification.

You may find many more **unauthorised modifications** inside when you open up the equipment on a **Detailed** Inspection.

# Visual/Close Inspection
## Bolts, Cable Entry Devices (Direct and Indirect) and Blanking Elements are of the Correct Type Complete and Tight.

Code A11

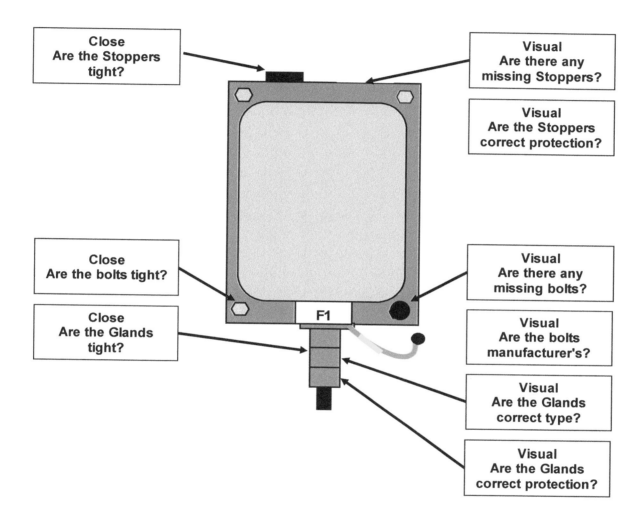

Most enclosures will be minimum IP54. If bolts, glands, stoppers etc. are slack then this IP protection would be compromised, especially if the item was a restricted breathing light fitting. Inspectors should be looking at the points above on **Visual** or **Close** Inspections. When looking at gland type Inspectors can only see on a **Visual** Inspection if the gland is compression or SWA. If the gland is required to be a barrier gland then this could only be established on a **Close** Inspection.

# Visual Inspection
## Are All of the Cover Bolts Present & Manufacturer Design?

### Introduction:

We take the nuts & bolts in Hazardous Area Equipment totally for granted. We do not look at them as actually part of the 'Protection' itself and something that has been tested by a notified body. We do not look at the fact that wrong bolts, however new or strong they appear, may not have the tensile strength to prevent the propagation of a catastrophic equipment explosion. Missing bolts can leave gaps in the equipment where gas, dust and water can impregnate the equipment and cause serious problems.

### Any Missing Bolts in Bulkhead Lighting?

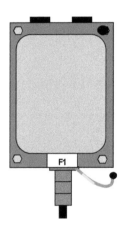

Missing bolts in **Exe** 'increased safety', **Exn** 'reduced risk' or **Ext** protection by 'enclosure', could jeopardise the IP54 minimum ingress protection or affect a restricted breathing fitting. **Remember this equipment is not made to withstand an explosion.**

Again **Exi** 'intrinsically safe' equipment is IP54 minimum seal so although no risk of explosion, water inside the equipment can have very traumatic effects on equipment such as printed circuits.

### Any Incorrect Bolts/Screws?

If a single bolt is odd, the chances are that this is not a manufacturer's bolt. Some situations are not as obvious as the diagram right. It is most important that the bolts are correct as they play an important part of the protection. Roofing bolts, by coincidence, usually fit although obviously they should not be used.

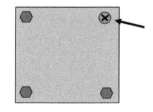

The bolts are tested by the notified body, such as **BASEEFA,** as part of the equipment and could play a very important part in withstanding the stresses produced by an explosion in, say, Exd flameproof equipment. The wrong bolts may fail and allow the equipment to blow apart. Remember Exd flameproof equipment requires special tools to remove the cover e.g. the grub-screw should only be able to be removed with a hexagonal key. If the grub screw has, say, a flat screwdriver head it can only be the wrong screw and not the manufacturer's product.

# Visual Inspection
## Are All of the Cover Bolts Present & Manufacturer Design? (Exd Flameproof Equipment)

## Code A11-2

### Any Missing Bolts in Flameproof Covers?

Missing bolts in Exd flameproof equipment could leave the flameproof gap larger than it should be. In this case gas or vapour could impregnate the fitting and any resulting explosion could be allowed out. This of course could have a catastrophic effect on the plant.

### Any Missing Grub Screws?

If the equipment has a screw lid then usually there is some kind of locking device to ensure that the cover does not vibrate loose. This may include grub-screws in the bottom of the threaded covers. Check that the grub screw is present. Replacement grub screws must be of manufacturer's design. (Hexagonal key.)

### Any Cover/Grub Screw in the wrong position?

The diagram (right) shows the position of the grub-screw. These screws fit into indents in the face of the cover (which is not a flameproof face). These indents are usually at 6 o'clock & 12 o'clock. If the grub screw is not at these positions then the likelihood is that the lid has been overtightened or in very disastrous cases is loosening up. In this type of box the thread of the lid itself is the flameproof path and the rule of 5 threads must be followed.

If the lid was to loosen up then there may not be enough threads left to form the flameproof path should an explosion occur inside of the equipment.

# Visual Inspection
## Are All of the Cover Bolts Present & Manufacturer Design? (Certified Boxes for IS)

### Code A11-3

### Certified IS Junction Boxes?

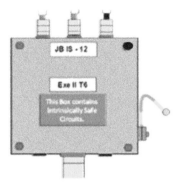

In the IS world there can, of course, be no arcs or sparks to cause any problems but that does not make the issue of any missing bolts or screws on the enclosures any less important.

IS multi-loop junction boxes can be like the model to the left which has 4 bolts to open it. If there is more than one loop in the box then the box must be Certified and any bolts would be manufacturer's bolts. Stoppers or glands would also have to be Certified. If one of the bolts is missing then the IP of the box is in jeopardy.

What some Inspectors tend to forget is that these boxes have gland plates with around 8 screws in them. If even one is missing, this compromises the IP of the box.

### Non Certified IS Boxes?

If we look at the barrier enclosure to the right, this box is usually in a non-hazardous area. The box is classed as 'Simple Apparatus' and does not need to be Certified.

This does not mean that it is allowable for any bolts to be missing from the cover as, if so, dust might impregnate the interior and make it unsafe.

In a small set up, see-through boxes are handy for showing up any problems. A Complex of any size would have large panels with racks & racks of barriers.

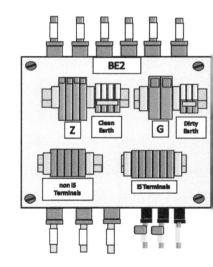

# Visual Inspection
## Are All of the Glands the Correct Protection & Complete? (Gland Protection)

Code A11-4

Introduction:

On a **Visual** Inspection, what are Inspectors looking for in the way of glands? Firstly are there any uncertified glands in Certified Equipment? Does anything look wrong with the gland? Is the gland 'Protection' correct for the protection of the equipment into which it is fitted, for instance, Exd glands in Exd equipment etc. All the Inspector has at the moment is an **Area Classification Drawing.**

Dates:

Let us go back to **before 2003**. At this time a **ONE SEAL UNCERTIFIED** gland could be used in Exe increased safety equipment, Exn reduced risk, and Exi intrinsic safety quite legally, and there will still be thousands out there in the field.

**From 2003 – 2008** it was legal in all of the above protections to use a **TWO SEAL UNCERTIFIED** gland, but a **ONE SEAL UNCERTIFIED** gland must no longer be used.

 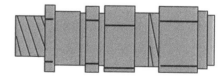

From **2008 to the present day**, the rule is that a correct **CERTIFIED** gland be used in **CERTIFIED** equipment. Universal glands are ideal. If the uncertified gland (Pre-2008) has to be removed from the equipment **FOR ANY REASON** e.g. rewiring or maintenance, **THEN A CERTIFIED GLAND MUST GO BACK**.

So to sum up gland protection, it must be a correct Certified gland in Certified equipment and on this **Visual** Inspection no tools may be used to move the gland, the protection markings must be read with the gland in place without turning it. When adaptors or reducers are used they also must be of the correct protection.

# Visual Inspection
## Are All of the Glands the Correct Protection & Complete? (Gland Protection)

### Code A11-5

#### Gland fitted incorrectly (SWA Showing):

The Inspector is looking for steel wire armour (SWA) protruding out the back-nut of the gland. If this is the case, the sheath seal will be ineffective. Water, gas or vapour can enter here and may bypass the deluge seal if one is fitted, leaving only the inner seal of the two seal gland to be effective.

This of course is usually bad workmanship and may not be quite as blatant as the diagram on the left so look carefully.

#### Gland fitted incorrectly (Seal Showing):

The sheath seal may be faulty and protruding out of the top of the gland as in the diagram on the right. This situation, like the instance above, will mean an ineffective sheath seal and gas, water or dust can then bypass this section of the gland so again you are relying purely on the inner seal. To sum this up, you only have a **ONE SEAL** gland in this case, as the sheath seal is ineffective.

#### Reducers fitted incorrectly:

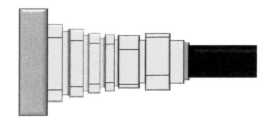

When reducers are used you can only use **ONE REDUCER PER GLAND** not a Christmas tree of them as shown left. There is a case that the volume could be increased with a large number of reducers.

# Visual Inspection
## Are All Stoppers Present and the Correct Protection?

Code A11-6

Introduction:

On a **Visual** Inspection the Inspector would check the equipment to see if any stoppers were missing. The next check should be to see if the stoppers that are in place are the correct 'Protection' for the equipment e.g. Exd flameproof stoppers in Exd flameproof equipment. All the Inspector would have at this point is an **Area Classification Drawing.**

Stopper Correct IP:

The stopper to the right is a brass/metal stopper with an 'O' ring so that when it is installed this ring helps to seal into the equipment. When the stopper is in place you may not see this 'O' ring.

Stopper Metals:

The stopper does not have to be brass. The stopper to the left is stainless steel.

Again you will notice the 'O' ring.

Stopper correct Protection:

The stopper **must** be of the correct 'Protection' for the equipment e.g. an Exd stopper in Exd equipment and an Exe stopper in Exe equipment.

Stoppers are usually screwed into the equipment from the outside, but it is possible to have stoppers screwed in from the inside. These are usually on equipment like bus chambers of lighting panels. This would force you to isolate before removing the stopper.

Black Plastic Stoppers:

In many cases, although not always, Exe stoppers are made of black plastic as shown left. Under **no** circumstances must these be used in Exd flameproof equipment. These stoppers have an 'O' ring to ensure the IP in Exe, Exn, Exi & Ext.

# Visual Inspection
## Are All Stoppers present and the Correct Protection?

### Code A11-7

### Stoppers fitted into Reducers:

 Stoppers must **never** be fitted and used in reducers. If there is a 25mm hole I suppose that it would look easy to get a 25mm – 20mm reducer and use a 20mm stopper. The correct size stopper must be used for the size of the hole.

### Transit Stoppers:

Transit stoppers are just what they say, they stop dirt and dust getting into the equipment in transit and while it sits on a stores shelf. These are usually very bright colours such as yellow or red. This temporary stopper can be screwed, a push fit or simply a sticky label over the hole. They must be removed and the correct Certified stopper used.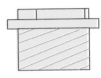

### Completely Parallel Stoppers:

 It is unlikely that there will be any of this type of stopper (left) still in use, but very old stoppers have been known to be parallel with no rim, so if you kept on tightening it would drop into the equipment. These would have to be fitted with a locknut to lock the stopper in place.

### IS Junction Boxes:

If a junction box is used in an IS loop and there is only one loop, then the junction box need **not** be Certified and classed as **'Simple Apparatus'** and can be a biscuit tin (so long as the IP is suitable). If the box has more than one loop then the box must be Certified e.g. Exe. The box then becomes not quite so simple as for one loop. If the box is Certified Exe then Exe glands & stoppers must be used.

### Certification:

 On a Visual Inspection you should be able to read the Certification with the stopper in place. If you cannot see any markings then you assume that it is not Certified.

# Close Inspection
## Are All of the Cover Bolts/Grub Screws Tight?

## Code A11-8

### Introduction:

When we completed the **Visual** Inspection it was a matter of whether the cover bolts/screws were present and of manufacturer's design. Now in a **Close** Inspection the Inspectors are looking to see if the screw lids, cover bolts and screws are all tight.

If the cover bolts or screws are not tight it could have a very detrimental effect on the protection of the enclosure. Let us take Exd flameproof equipment, if the bolts are loose the flameproof gap may be larger than it should be and Inspectors would be able to insert feeler gauges.

If the equipment is Exe, Exn or Ext protection by Enclosure then the IP property will be reduced and water, gas or dust could enter the equipment. All of these situations are, of course, unsafe.

### Is the Exd Threaded Cover Tight?

If the grub screw was found to be missing on the **Visual** inspection then it could be that the lid is loose. On the **Close** Inspection the Inspector can get hold of it and check it.

I have, as a young electrical technician, caught a lid on the last few threads so I can assure you it does happen.

### Are the Marshalling Box Lid Bolts tight?

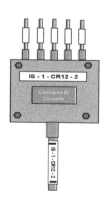

Just imagine if the lid bolts were loose on the lid of the marshalling box to the left. What would happen if water was to enter the enclosure?

Let me tell you: absolute devastation. It would start with various process systems either malfunctioning or giving peculiar readings, which in itself, would be very dangerous.

Finding the problem in these cases is not easy as there could be over 200 of these boxes scattered all over the plant.

That is why these Inspections are completed. Why were the bolts slack to start with? Did someone forget to tighten them up? Did the bolts work loose themselves and, if so, how?

# Close Inspection
## Are All of the Gland Sections Tight?
## Should the Gland be a Standard or Barrier?

### Code A11-9

### Introduction:

On a **Close** Inspection the Inspector can now get hold of the glands and see if they are loose. Inspectors can also determine if the gland should be a standard gland or a barrier gland because they have more documentation here.

### Gland Sections Loose:

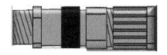

Ensure all gland sections are tight. Inspectors should not be able to undo the gland by hand, not even the knurled section.

### Barrier Glands:

Barrier glands would only be picked up on a **Close** Inspection as on a **Visual** Inspection there would not be any documentation available.

As you can see from the diagram above there is compound after the steel wire armour trap. This is to stop any explosions from going down the cable gaps in the middle and coming out on the weakest part of the sheath. Only usually used in certain circumstances, for instance if the equipment is flameproof with an internal source of ignition, e.g. a switch. Obviously barrier glands **are not** required on equipment such as junction boxes as these have no internal source of ignition. There are two policies at the moment as to when a barrier gland should be used:

Policy 1: An Exd flameproof/Gas Group llC item of equipment with an internal ignition source in a Gas Group llC Area:- **Barrier Gland**.

Policy 2: The IEC have determined that if the cable, gland back-nut to gland back-nut, is 3 Metres, or under, and the equipment is Exd flameproof with an internal ignition source, then a barrier gland is used on the end with internal ignition.

So what is there to inspect? Well mainly, is there a barrier gland where there should be one? As the Inspector is unable to touch the gland on a Visual Inspection.

### Gland Temperature Class:

Glands do not usually have a temperature class as such. Their temperature defaults from **-20C to +80C.** Just remember what a large portion of the gland is, i.e. rubber, when looking at very low or high temperatures.

# Close Inspection
## Are All Stoppers Tight?

## Code A11-10

### Introduction:

With a **Visual** Inspection the Inspector would be looking to see that the stopper is **present** and has the correct protection and is correct for the equipment. With a **Close** Inspection, the Inspectors are checking to see if the stopper is **tight.**

### Loose Stoppers Exd:

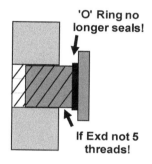

**'O' Ring no longer seals!**

**If Exd not 5 threads!**

Loose stoppers are a hazard and, depending upon the vibration level, could vibrate out of the equipment completely.

If the stopper is in Exd flameproof equipment, coming loose means it may not end up with the correct number of threads entered.

Very old stoppers have been known to be parallel with no rim, so if you kept on tightening it would drop into the equipment. These would have to be fitted with a locknut to lock the stopper in place. It is unlikley that you will come across these old stoppers.

### Loose Stoppers Exe, Exn & Exi:

If the stopper is in Exe increased safety, Exn reduced risk, Exi intrinsically safe or Ext protection by enclosure equipment, by coming loose it may not end up with the correct IP seal.

For this equipment the minimum IP is IP54 and unless the 'O' ring on the stopper is tight up against the equipment then the ingress protection of that equipment could be compromised and water, gas, vapour or dust could get in.

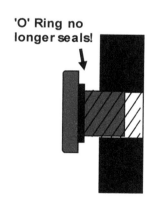

**'O' Ring no longer seals!**

### Loose Stoppers Restricted Breathing:

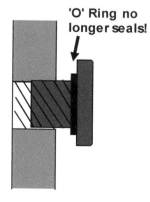

**'O' Ring no longer seals!**

If the stopper is in ExnR restricted breathing equipment, by coming loose it may not end up with the correct IP seal.

For this equipment the minimum IP is IP54 and unless the 'O' ring on the stopper is tight up against the equipment then the ingress protection of that equipment could be compromised and water, gas or vapour could get in.

The restricted breathing of the fitting would be non-existent.

# Visual Inspection
## Are Threaded Covers correct Type & Secure?

## Code A12

### Introduction:

Threaded covers in many cases have grub screws to stop them from becoming loose. Sometimes these screws, being so small, go missing during maintenance or simply vibrate out if they are loose. The threaded cover can vibrate loose and in extreme cases vibrate completely out. On a **Visual** Inspection the Inspectors would not know if the lid is loose unless they can see thread.

### Any Missing Grub Screws?

If the equipment has a screw lid then usually there is some kind of locking device to ensure that the cover does not vibrate loose. This may include grub-screws in the bottom of the threaded covers. Check that the grub screw is present. Replacement must be a manufacturer's product.

Remember Exd flameproof equipment requires special tools to remove the cover i.e. the grub-screw should only be able to be removed with a hexagonal key. If the grub screw has, say, a flat screwdriver head it **must** be the wrong screw and not the manufacturers.

### Any Cover/Grub Screw in the wrong position?

The diagram (right) shows the position of the grub-screw. These screws fit into indents in the face of the cover (not a flameproof face). These indents are usually at 6 or 12 o'clock. If the grub screw is not at these positions then the likelihood is that the lid has been overtightened or in a very disastrous case is loosening up. With a **Visual** inspection you would not know if it was loose.

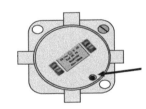

In this type of box the thread of the lid itself is the flameproof path and the rule of 5 threads must be followed. If the lid was to loosen up then there may not be enough threads left to form the flameproof path should an explosion occur inside of the equipment.

### Any Screw Cover in IS Equipment?

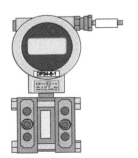

Instruments like the one in the diagram on the left usually have a screw lid. In many cases the lid has special notches so that a tool can be inserted to open the cover.

Any damage to the lid or if it does not look as if it fits correctly should be reported. Usually the instrument would be IS so no problem with sparking or explosions.

# Close Inspection
## Are Threaded Covers correct Type & Secure?

### Introduction:

Threaded covers in many cases have grub screws to stop them from becoming loose. Sometimes these screws, being so small, go missing during maintenance or simply vibrate out if they are loose. The threaded cover can vibrate loose and in extreme cases vibrate completely out. Now on the **Close** Inspection the Inspectors can get hold of the lid to check if it is loose.

### Any Missing Grub Screws?

If the equipment has a screw lid then usually there is some kind of locking device to ensure that the cover has not vibrated loose. Get hold of the cover and see if it will turn. Is the grub-screw in the bottom of the threaded cover tight? Remember if Inspectors identify that the grub screw is in fact missing they can now on this **Close** inspection get hold of the lid to see if it is loose.

### Any Cover/Grub Screw in the wrong position?

The diagram (right) shows the position of the grub-screw. These screws fit into indents in the face of the cover (not a flameproof face). These indents are usually at 6 o'clock or 12 o'clock. If the grub screw is not at these positions then the likelihood is that the lid has been overtightened or in a very disastrous case is loosening up.

In this type of box the thread of the lid itself is the flameproof path and the rule of 5 threads must be followed. Again on this **Close** Inspection the Inspectors can get hold of the lid and check if it is loose.

### Any Screw Cover in IS Equipment?

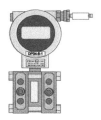

Instruments like the one in the diagram on the left usually have a screw lid. In many cases the lid has special notches so that a tool can be inserted to open the cover.

Inspectors can get hold of the lid to see if it is loose.

# Detailed Inspection
## Are Exd Joint Faces Clean and Undamaged?

## Code A13

### Introduction:

Electrical isolation required and Gas Free Certificate. On a **Visual** Inspection the outside of the equipment is all that can be checked for damage, but now that the equipment has been opened up on the **Detailed** Inspection, flange faces can be inspected. I have included screw threads of the enclosure to be checked because these, of course, are flameproof paths.

### Are all Exd Flameproof Faces Clean & Undamaged?

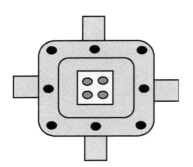

Ensure that all of the flameproof flange faces are clean and free from damage. **DO NOT FORGET THE EQUIPMENT LID.** Even the slightest scratch or mark should be reported.

On closing the enclosure ensure that a smear of non-setting grease is applied to the flange faces on both the equipment and the lid, all bolts should be 'Spanner Tight' and the flameproof gap checked

### Are all Exd Threads Clean & Undamaged?

Ensure that all of the flameproof threads are clean and free from damage. **DO NOT FORGET THE EQUIPMENT LID.** Even the slightest mark should be reported.

On closing the enclosure ensure that a smear of non-setting grease is applied to the threads on both the equipment and the lid. The lid should be screwed back on with the hexagonal grub screw in the right position (usually 6 o'clock or 12 o'clock) and tightened 'Spanner Tight'.

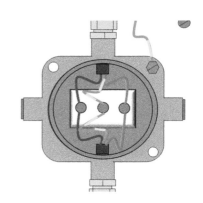

# Detailed Inspection
## Is Condition of Exd Enclosure Gasket Satisfactory

## Code A13-1

Introduction:

Electrical isolation required and Gas Free Certificate. On a **Visual** Inspection the outside of the equipment is all that can be checked for damage, but now that the equipment has been opened up on the **Detailed** Inspection, any gaskets and the inside of the enclosure can be checked.

### Is the Enclosure Gasket satisfactory?

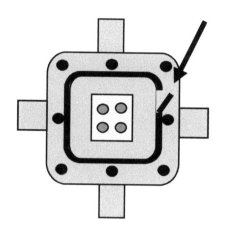

Many flameproof enclosures these days have gaskets or 'O' rings. These can become damaged over time and break or fray. The problem here of course is that bits of gasket can get onto the flameproof face and cause a gap.

A manufacturer's gasket should be obtained and fitted. Do not leave the gasket off if the equipment is meant to have one.

Some manufacturers of flameproof bulkhead light fittings these days design gaskets or 'O' rings into their flameproof enclosures to give extra IP. These gaskets are sometimes made of a very soft, almost fabric, material and can easily become damaged over time and break or fray. The problem here of course, as above, is that bits of gasket can get onto the flameproof face and cause a gap.

A manufacturer's gasket should be obtained and fitted. Do not leave the gasket off if the equipment is meant to have one.

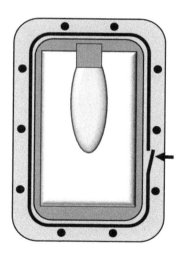

# Detailed Inspection
## Is Condition of Enclosure Gasket Satisfactory?
## (Exe, Exn & Exi equipment)

## Code A14

### Introduction:

Electrical isolation required and Gas Free Certificate. On a **Visual** Inspection the outside of the equipment is all that can be checked for damage, but now that the equipment has been opened up on the **Detailed** Inspection, gaskets can be checked. If the gasket is faulty then water, dust, gases & vapours may be able to enter the enclosure and set up a very dangerous situation.

### Is the Exe Increased Safety Enclosure Gasket satisfactory?

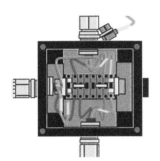

Sometimes when Inspections are carried out people remove the enclosure lid and put it on the floor whilst they inspect the enclosure.

On completion of their Inspection they pick up the enclosure lid and put it back on and not once do they inspect the lid itself to see if the gasket is intact or if anything is amiss.

### Is the Exe or Intrinsic Safety Enclosure Gasket satisfactory?

Usually with the type of enclosure shown on the right the gasket is made of a rubber type material and fastened to the lid itself.

Any parts of the gasket that are damaged or missing will definitely compromise the IP of the box. An Exe box would have to maintain something like IP54 minimum.

If it is an IS system box there is no problem with sparking, however, water could cause the IS loop to shutdown or give false readings.

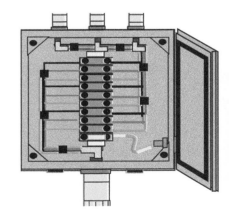

# Detailed Inspection
## Is there Any Evidence of Water or Dust Inside the Enclosure?

## Code A15

### Introduction:

Electrical isolation required and Gas Free Certificate. When on a **Detailed** Inspection, the cover is removed from the equipment we now must see if there is any water or dust inside. If there is, then the equipment IP is not good enough. To get water into the equipment the IP must be less than IP54. To get dust inside the enclosure the IP must be less than IP5X or IP6X. This could of course all be due to a broken or faulty gasket or 'O' ring. Check the lid gasket in this case. Exd flameproof equipment is not of course waterproof although non-setting grease will go a long way towards rectifying this.

### Check the Seal:

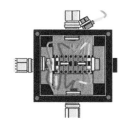

The seal on the box is extremely important as being the one thing that prevents gas, vapour, dust or even water from entering. The equipment, whatever it may be, is manufactured to a certain IP and in many cases IP54 is the minimum. If we take a junction box as on the right, the seal is in the lid and is either a flat fibre type or an 'O' ring. If this is damaged or missing the box is no longer protected.

If the seal/gasket seems to be intact then the water must be getting into the equipment some other way. Try looking at the gland seals to see if they are intact. Are there any cracks in the equipment casing or flaws in the cable sheath exposing SWA?

If the equipment is Exd Flameproof try putting an extra smear of non-setting grease onto the flange face.

### Marshalling Boxes:

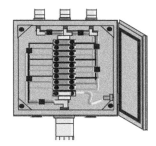

In equipment similar to the example on the left, the seal is usually made of a rubber type material and is fastened on the door itself. Ensure it is not damaged in any way otherwise the same dangers apply that were discussed in the above equipment. Check the seal and ensure that there is no water or rust in the bottom of the equipment.

I have known gland plates to completely rust off clear of the equipment before today, which could be extremely dangerous depending upon the duty of the junction box.

# Close Inspection
## Dimensions of Flange Joint Gaps Within Maximum Values Permitted?

Introduction:

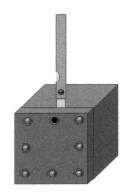

The flameproof gap must be checked after bolting up an Exd flanged piece of equipment. Bolts missing may result in feeler gauges entering (see left).

The flameproof gap should be **0.15mm** for Gas Group llA & llB flanged piece of equipment and **0.1mm** for flanged Gas Group llC with the enclosure volume <500 cubic cm.

If the volume of the equipment was to exceed the above volume of 500 cubic cm then the gap tolerance could go down to **0.04mm.**

Flanged llC equipment is not as common as llA & llB and spigot & screw flameproof equipment is usually preferred for Zone 1 areas because they are more effective.

Short feeler Gauge:

Short

**DO NOT USE A LONG FEELER GAUGE** as the equipment could be switched on when doing a close inspection and if you insert the feeler gauge you do not want it to protrude inside of the equipment and perhaps touch something that is live.

Screw Threads:

Although this part of the Standard does not specifically apply to anything flameproof that screws into equipment, I think it is worth mentioning in this section of dimensions that if the item is a screwed lid, gland, stopper or conduit that screws into a flameproof enclosure that **5 FULL THREADS** have to be entered to be an effective flameproof path.

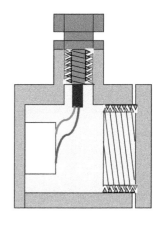

The equipment must be **6 FULL THREADS** deep or **8mm axial length** to allow you to get your 5 in.

# Detailed Inspection
## Are All of the Electrical Connections Tight Inside?

## Code A17

### Introduction:

Electrical isolation required and Gas Free Certificate. Now that the cover/lid has been removed the screws on the inside of the equipment must be checked for tightness. Remember it might be years before another **Detailed** Inspection is done. Slack electrical connections can cause high resistance and heat.

### Terminal Screws:

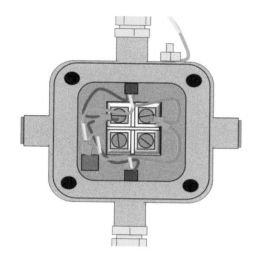

As well as the main screws holding cables in the terminal block, any spare screw terminals are also tightened at this stage to stop the screws falling out due to vibration.

If any screws clamping cables are loose of course, depending upon what the equipment is, this could cause heat. If the equipment is Exe increased safety or Exn reduced risk, heat is the last thing required.

### Marshalling Boxes:

Admittedly being intrinsically safe there would be no problem with heat or sparks, but loose connections may well affect the IS loop and of course the running of the plant.

Again although an intrinsically safe system, it is best to complete this exercise with the loop dead in case it affects the running of the plant.

Remember company policy may insist on dead maintenance regardless of IS.

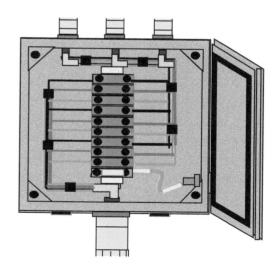

# Detailed Inspection
## Are All Unused Terminals Tight?

## Code A18

### Introduction:

Electrical isolation required and Gas Free Certificate. As mentioned in A17, all unused terminal screws must be tightened to ensure that they do not work loose, fall out and cause a spark. In Exe & Exn especially, the manufacturers go to great lengths to ensure that when you connect up cables in an enclosure, there can be **NO CHANCE AT ALL** of anything causing heat or sparks. This is one of their precautions.

This task must be completed now as it might be several years before anyone returns to carry out another **Detailed** Inspection.

### Marshalling Boxes:

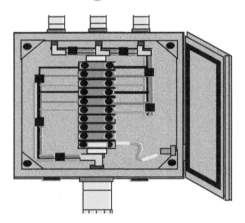

Admittedly being intrinsically safe there would be no problem with heat or sparks, but loose connections may well affect the IS loop and of course the running of the plant.

Again although an intrinsically safe system, it is best to complete this exercise with the loop dead in case it affects the running of the plant.

Remember company policy may insist on dead maintenance regardless of IS.

### Other types of Connection Block:

Now the terminal screws in this type of Exe or Exn enclosure are much more substantial. If these screws were to come loose and fall out there would be a slim possibility that they could cause a problem. It is unlikely that a JB of this kind would be IS. Spare screws in this case must be tightened to prevent any problems.

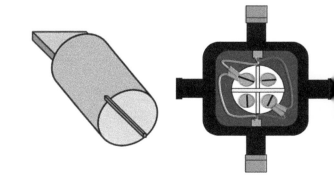

# Detailed Inspection
## Are Exn Enclosed Break Devices Undamaged?

## Code A19

### Introduction:

Electrical isolation required and Gas Free Certificate. **The Enclosed Break** definition is similar to that of flameproof i.e. there is an enclosed chamber where the arcs & sparks take place, but internal ignition will not propagate to the outside. This equipment, of course, relies mainly on the sealing of the equipment and has no flame-path that you can check with feeler gauges.

### Enclosed Break Lamp-holders:

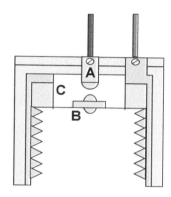

One nC component that immediately springs to mind is the enclosed break lamp-holder. Live maintenance should not be completed but if you look at the lamp-holder on the left there is a chamber 'C' which is the enclosed break, there is a contact 'B' where the cap of the lamp will make contact first and contact 'A' where the mains supply comes into the lamp-holder. If someone were to screw a lamp into the Edison screw, when the cap of the lamp touched 'B' it would not light. If you kept on screwing the lamp it would push 'B' onto 'A' and then the lamp would light. Any arcs & sparks would be sealed in chamber 'C'.

### Misc. Information:

These lamp holders are fitted in Exn fittings which may be ExnR (restricted breathing) which can only be installed into a Zone 2 area where gas or vapour is not expected to be present.

These lamp holders have to be made in such a way that any ignition inside cannot be transmitted to any atmosphere outside. This of course is similar in description to 'Flameproof' except that in this case there are no flameproof faces.

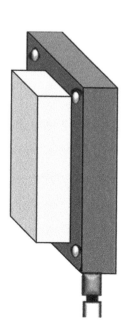

The sparking section of the lamp holder must be able to withstand the pressure of an explosion if gas was to penetrate the enclosure.

The enclosure has a volume limitation of not more than 20 cubic centimetres.

# Detailed Inspection
## Are Hermetically Sealed Devices Undamaged?

## Code A19-1

### Introduction:

**Hermetically Sealed** equipment such as multi-pin plug-in relays, solenoids etc. The equipment comes under the 'Protection' title of ExnC.

### Hermetically sealed Equipment:

What is a hermetic seal and how do we check it? Hermetic seals are absolutely airtight, they are the opposite of flameproof in that they prevent invasion of outside contaminates e.g. water, dust, gas etc.

The most common types of hermetic seals are probably double glazing and, in the food industry, tins of food with the pull ring widget on the top.

Electronic & electrical equipment can also be hermetically sealed e.g. solenoids, 8/16 Pin plug-in relays (right), and light bulbs for instance as they can form a perfect glass to metal seal in the light bulb, keeping the gas in and the outside atmosphere out.

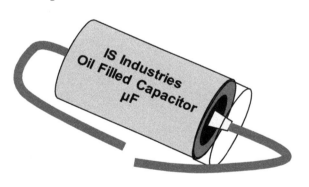

Another hermetically sealed item of equipment might be an oil filled capacitor as in the picture to the left. So obviously if there were signs of oil around the unit then more investigation must be carried out. If the hermetic seal is broken and oil allowed to escape or air enters into the item there could be an explosion.

Resin filled capacitors can be obtained which are also hermetically sealed.

As mentioned above, probably one of the most common examples of hermetic sealing is the glass to metal part of a light bulb. The metal would have the same coefficient of thermal expansion as the glass so the seal cannot be broken by heat expanding the different materials.

# Detailed Inspection
## Are Exn Encapsulated Devices Undamaged?

## Code A20

## Introduction:

Electrical isolation required and Gas Free Certificate. When we look at encapsulation there are two names that spring to mind - **'Potting'** and **'Encapsulation'**. In **encapsulation** the flying tails are wiring that is connected onto the electronics first, then it is dipped in resin.

**Potting** is where the flying tails are put on and then a mould is made to take the electronics and resin poured in to seal the whole thing.

During the setting process the resin can reach temperatures of +200C so the manufacturers must consider that the components & wiring may also reach this temperature.

## Damage:

Are any of the encapsulated components damaged inside of the Exn reduced risk enclosures, which will include moisture damage should the seal have become faulty?

## Encapsulated Components:

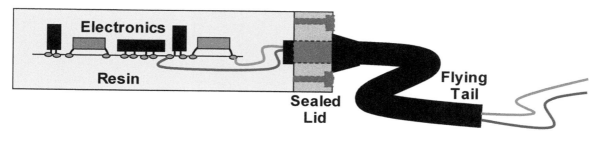

With encapsulated, Inspectors are looking at a chamber which is airtight, for instance equipment such as electronics with higher mains current. Flying tails are soldered onto the contacts of the component during construction and then an epoxy resin is packed into the chamber and a lid put on, never to be removed again and the component will come complete with its supply cable. This equipment also comes under the **'Protection'** heading of **'ExnC'**.

Here we are looking at ExnC components which are fitted into Exn equipment which, of course, is only for Zone 2 and not as highly Certified as Exm encapsulation.

# Detailed Inspection
## Are all Flameproof Components Undamaged?
## (In Exe & Exn Enclosures)

## Code A21

### Introduction:

Electrical isolation required and Gas Free Certificate. This is usual where the enclosure is Multi-Certified. Sometimes when you pick up an Exde light switch, the Exe section is usually the enclosure itself i.e. made out of plastic, the Exd section is the internal switch mechanism, not because of flameproof paths, but there is no volume there and the chamber will be made to withstand an explosion from within.

The internal switch mechanism will most likely have a 'U' after its Atex number denoting that it is a **'component part'** of that unit, so the mechanism cannot be used by itself or in some totally different equipment.

### Multi-Certified Unit:

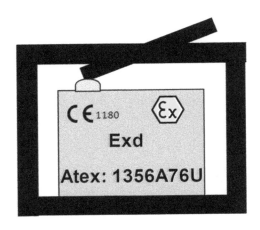

The Multi-Certified enclosure on the left is a typical Exde switch unit where the outside is plastic i.e. Exe increased safety, and the inside mechanism is Exd flameproof. As mentioned above, the Exd enclosure, because it has a 'U' after the Atex number, means that it cannot be used independently of the switch enclosure. You can fit an identical replacement should it develop a fault.

The diagram to the right is a representation of a 240 volt Exde socket outlet of the type where the plug is turned to switch it on and off. The outside case is black plastic so that is the Exe section. Inside this enclosure is a switch which is Exd, not because it has flameproof gaps but because there is no volume there. This switch would again have a 'U' after the Atex number denoting a **'Component Part'**, and cannot be used in anything else except an identical unit. This switch unit has a blue cap denoting 240 volts, they can also be obtained at 415 volts with a red cap and 110 volts with a yellow cap.

# Detailed Inspection
## Exn Restricted Breathing Enclosure is Satisfactory?
## Exn Test Port if fitted is Functional?
## Breathing Operation is Satisfactory?

## Codes A22, A23, A24

### Introduction:

Electrical isolation required and Gas Free Certificate. ExnR restricted breathing enclosures are constructed with a high level of sealing where if there was to be a long term concentration of gas outside of the enclosure, which would be unlikely as these enclosures can only go into Zone 2 areas, inevitably limits the entry of a gas to below its lower flammable limit. Many bulkhead light fittings are restricted breathing, but it can apply to other enclosures that are not lighting. The seal must be checked to see that it is in good condition once the enclosure is opened up. It does not, as many people think, completely seal. Every so often, especially if maintenance has been completed, the enclosure must be tested with specialised test equipment to ensure that the restricted breathing condition still exists.

### Test Port:

Many enclosures are fitted with a test port in the housing to allow test equipment to be fitted to the enclosure to test that the restricted breathing property of the enclosure is still satisfactory. This port is sealed when the enclosure is in normal use by a correct manufacturer's stopper. This test port, if fitted, must be inspected to ensure that it is in working condition.

### Test Kit:

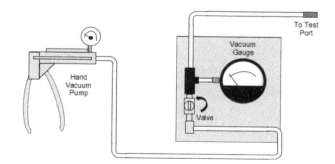

The test kit is a hand operated vacuum pump which is connected to the test port on the enclosure (if fitted). This in turn will create a vacuum inside of the enclosure to manufacturer's instructions. This may be a routine test or take place after maintenance.

Engineers may not wish to carry out this test too often as it may take around 10 – 15 minutes per fitting.

# Close Inspection
## Are Breathing & Draining Devices Tight?

Code A25

Introduction:

On a **Visual** Inspection you are checking that the draining device is the correct protection for the equipment. On a **Close** Inspection you are checking that it is fitted correctly and is tight.

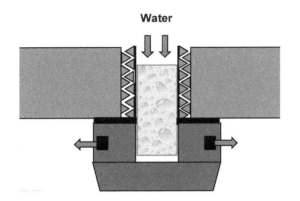

**Water**

These draining devices are Atex approved and can be obtained for Exd & Exe equipment.

As you can see by the diagram the water drains from the equipment into a filter arrangement where it is safely let out without disturbing the protection of the equipment.

Manufacturers should really be consulted before fitting one.

Desiccators:

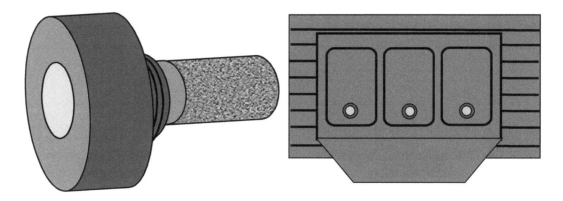

I have included this desiccator in this section although unless you are inspecting a HV motor it is unlikely that you will come across them. I could have included it in the dry Insulation section. As you well know on 3.3kV, 6.6kV & 11kV motors the terminal block must be segregated (separate chambers for each of the three phases). In each chamber there is likely to be what is called a desiccator. They are a type of breather and absorb moisture from the air.

These desiccators are usually filled with a blue silica gel which turns pink when damp. If pink they then need replacing. Molecular sieve is another type. They should only be replaced on a stopped HV motor **NOT** whilst it is running.

# Visual inspection
## Are Any Fluorescent Lamps Displaying 'End of Life' Effects?

Code A26

### Introduction:

It is very important that any lamps showing **'End of Life Effects'** are picked up. Fluorescent lighting is easily detected when nearing 'end of life' as the tubes start to blacken at either end. Sometimes the problem is that if the lamp was not to strike, the control gear could still be trying to make it go, so that this unit also burns out.

### Can you see at this stage if the Tube/Lamp is at 'End of Life'?

Fluorescent tubes that have reached their 'end of life' (EOL) status emit just a glow and are black on either end. The tube emitter has reached evaporation exhaustion and the debris has collected on the inside of the glass at either end (the black). As mentioned above, the problem here is that the control gear, i.e. the choke/ballast unit, could still be trying to make it fire and could burn out itself. There will be no separate starter unit.

Another problem is that the tube could overheat around the electrode area and, I have known in extreme circumstances, burn a hole in the glass tube. There are protective measures to prevent the choke/ballast causing overheating, called a **'Stop'** in the electronic circuit. IEC 61195 is the Standard.

### Mono-pin Tubes:

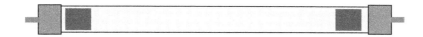

In older Exe fluorescent fittings the lamp may only have one pin as shown above. This is called a 'mono-pin' or cold cathode (no heaters) and in this case at end of life there will not be a glow at either end, it just will not work. There may still be black debris at either end.

Also, if the tube has internal powder missing in some parts the light efficiency will go down.

# Visual inspection
## HID Lamps Displaying 'End of Life' Effects?

Code A27

Introduction:

When we talk about high intensity discharge lamps or **HID lamps** we include the lamps 1 – 4 below. Each of these has its own problems. We talk about a lamp having an **'Efficacy'** which is mainly how much light is emitted for the power consumption of the lamp. Other factors also come into the equation when deciding what type of lamps to purchase for a large complex, such as how many thousand hours will the lamps light for, cost, environmental friendly gas, light pollution etc.

The High Intensity Discharge (HID) Lamp:

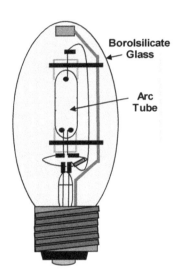

1   **High Pressure Sodium:** Yellowish light, contains mercury so lamps that do not work must be disposed of correctly. Restrike time in loss of voltage. Efficacy good. (How many Watts/Lumen Output.)

2   **Low Pressure Sodium:** Not allowed in hazardous areas. Contains what is called sodium metal in its internal 'U' Tube. If it is a damp day and the lamp was dropped, the moment the sodium metal came into contact with the damp air it would explode. (Efficacy not as good as the other three mentioned.) Used to be in street lamps.

3   **Metal Halide:** Similar to mercury vapour except the addition of metal halide gives the lamp a more whitish light than bluish. Efficacy good. (How many Watts/Lumen Output.)

4   **Mercury Vapour:** Used to be used in abundance in factories. Bluish light and again contains mercury so lamps that do not work must be disposed of correctly. Restrike time in loss of voltage. Efficacy good. (How many Watts/Lumen Output.) Borosilicate glass contains boron trioxide and has a very low coefficient expansion meaning it will withstand very high temperatures without cracking.

HID Lamps at 'End of Life':

Sometimes at **'End of Life'** these lamps go through a process called **'Cycling'** which is that the lamp starts and an arc forms in the arc tube but, as the lamp heats up and starts to get brighter it flashes out and the whole process starts again. It is to do with start voltage compared with how many volts required to keep the lamp lit. The colour of the light output also changes as the lamp gets older and its efficiency changes. Sometimes the HID lamps explode. Just remember what gases would be released, one of them being mercury vapour.

# Detailed Inspection
## Are All Lamp Ratings & Pin Configurations Correct?

## Code A28

### Introduction:

Electrical isolation required and Gas Free Certificate. When we talk about this subject we have to look at what lamp has been installed into the fitting? Sometimes there is a **'U' (Component Part)** or an **'X' (Special Conditions)** involved where a specific lamp is supplied with the fitting by the manufacturer and that is the one that has to be used.

### Conductive Coating:

There are lamps on the market where the Edison screw, for instance, is made of ceramic or very dense plastic with a conductive coating. These must not be used in a Certified fitting unless they came with, and have been tested with, the fitting.

You can change the conductive coated screw lamp for another if it came with the fitting, in which case it may have a **'U'** as a 'component part'.

### Emergency Lighting:

Emergency lighting must never have a lamp installed that has a 'Striking Time' such as mercury blended or sodium. The reason for this is that if the plant had several power cuts in quick succession, the emergency lighting would be off until the lamp struck, which could take several seconds to a minute before it reached full brightness leaving the plant in pitch darkness until the lamps decided to strike.

### Maximum Lamp:

In my day as a young electrical technician, many fitting data plates stated what size lamp had to be installed in that particular fitting. These days the manufacturers usually state 'Maximum' lamp in which case any lamp can be used up to that size.

I must emphasise here that if a manufacturer states that a particular lamp must be used then that is the lamp that must be installed.

# Visual Inspection
## Is there Any Damage to the Electric Motor?

## Codes A29

### Introduction:

The most common motor in a chemical factory is a 3 phase (electric power) squirrel cage (rotor looks like a squirrel/hamster's wheel) induction motor (power is induced into the rotor by the rotating magnetic field inside the motor). These motors are usually made out of cast alloy which sometimes suffers damage.

### Has the Motor got damaged Feet?

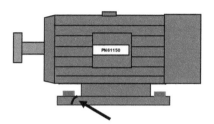

Look very carefully at equipment such as Exd motors especially where there might be some vibration.

The motor case is usually cast alloy and there may be damage such as cracked foot mountings.

### Is there any damage to the Cowl Mesh?

Check the motor fan cowl for damage to the mesh. Any holes, of course, could be dangerous as the rotating fan can then be touched with hands. When I was a young electrical technician we used to test the mesh holes with a **'British Standard Finger'** and if the mesh was too large then we would remove the cowl, chisel out the old mesh and change it for a smaller mesh.

The above is something you are not allowed to do today on a Certified motor. You have no idea what the airflow would be with a smaller mesh.

### Is there any damage to the Motor Cooling Ribs?

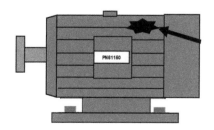

On the main body of the motor are cooling ribs or fins. Air is drawn in by the fan at the non-drive end of the motor and blown along the grooves made by the ribs. The grooves provide more surface area for cooling as well as providing a path for the airflow.

To a certain extent the ribs will be self-cleaned by the airflow, but debris can clog up the ribs such as pipe lagging which has fallen onto the motor. These ribs can also be damaged if anything heavy falls onto the motor.

# Visual Inspection
## Is the Motor Air Flow Impeded?

## Codes A30

### Introduction:

There are devices that can be installed into the motor windings to cut the power if the motor windings were to heat up. These devices are called 'thermistors'. In very large HV motors there might be other devices called resistance thermometers. Impeded airflow will heat up the motor. Most small induction motors will not have wired thermistors.

### Is the Motor Air Flow unimpeded?

In most companies their small, 415 volt, three phase motors depend purely on the overload unit tripping the control circuit if the motor was to go into overload.

As mentioned above there is no thermal cut out on these small 415 volt motors if the windings were to overheat due to, say, the fan cowl being blocked cutting down the airflow over the motor.

### Thermistor Cut Outs?

When the motor leaves the manufacturers sometimes these thermal cut out devices are installed, usually three of them in a small 415 volt motor, one in each phase winding, that can be put into the windings, as above, called thermistors and these devices detect temperature rise in the windings.

The problem here is cost. For what little use a thermistor is, an extra cable has to be run to the starter control from the motor to connect these thermistors into the control circuit. So for every motor on plant an extra cable besides the motor load cable has to be run, therefore companies will only bother with this on large, high voltage motors, not for small motors, because of the cost.

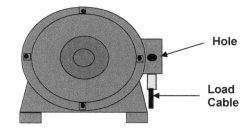

Of course an extra cable requires an extra hole in the motor terminal block and because the thermistors are not connected in, this hole is sometimes left open with no stopper.

# Visual Inspection
## Is the Motor Lifting Facility Satisfactory? (Company Policy.)

### Codes A30-1

### Eye Bolts & Shackles:

When the motor leaves the manufacturers an eye bolt is installed, usually in the top of the motor. A crane team can then put a **shackle** through the **eye bolt** connecting it to their lifting hook and chains and lift the motor into place.

If the eye bolt is not removed it could sit there for years before being required to lift the motor. Of course if the eye bolt is left in position on the motor it will not have been tested in all of that time.

**Eye Bolt**       **Shackle**

### Is the Lifting Eye still in the Motor?
### (Company Policy)

We are talking here about the eye bolt in the top, or sometimes the sides, of equipment such as an electric motor.

These are sometimes referred to as **'lifting eyes'** or 'dynamo eye bolts' and are not always round as the example above and to the right.

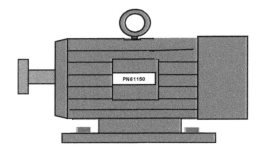

Leaving the lifting eye bolt in on a motor, although not exactly damaging in itself means that it can fail when the motor is being lifted if it has corroded over time. If the eye was removed and a bolt greased and put into the hole, a tested eye bolt would be screwed onto the motor when it had to be lifted.

At BP where I worked, the lifting eye bolt had to be removed from the motor and replaced with a greased, standard bolt. The eye bolt was then taken back to the workshop where it was tested, colour coded, labelled as to thread size and put on a rack until required. Sometimes on a larger heavier motor there may be several eye bolt points.

# Detailed Inspection
## Electric Motor Winding Connections

Code A31

Introduction:

Electrical isolation required and Gas Free Certificate. When the motor terminal block is opened for a **Detailed** Inspection there are several tests that can be done to ensure that the motor is safe and not heading for a major breakdown.

Motor Windings in Delta:

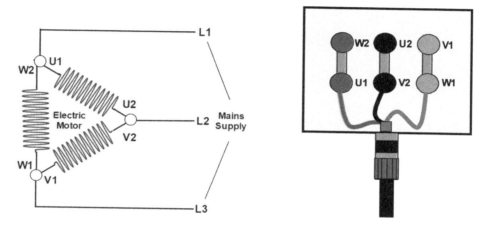

As can be seen from the above diagrams the electric motor in this case is connected in **'Delta'**.

Motor Windings in Star:

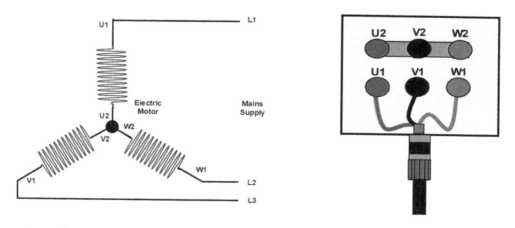

As can be seen from the above diagrams the electric motor in this case is connected in **'Star'**.

# Detailed Inspection
## Electric Motor Winding IR

## Code A31-1

Motor Insulation Resistance (IR):

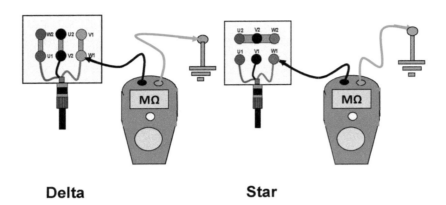

**Delta**                                  **Star**

Connect the Insulation Resistance Tester (Megger) as per the diagram above. Select **MΩ** and carry out the test. By all means try all three phases. (It will of course be the same.)

So what reading are we expecting here? Well I would say that a normal good reading would be anything from **10 to 100MΩ.** Anything under **5MΩ** needs investigation.

The seal on some motors may only be something like IP54, and if the motor has been subjected to wet conditions, water may have penetrated and caused rust in the terminal box which would affect the reading. I have opened up many motors and found the terminal box to be extremely rusty. If the motor has not run for some considerable time this might also adversely affect the reading.

## Insulation Resistance Readings:

When taking the Insulation Resistance Readings (IR) as above, what readings might be good/acceptable and what readings might be bad and require investigation? Remember to check company policy here with the electrical engineer.

| | | |
|---|---|---|
| 2 MΩ or Less | Extremely Bad | Motor may require removal |
| 2 – 5MΩ | Critical | Motor may require removal |
| 5 – 10MΩ | Abnormal | Engineer to Risk Assess |
| 10 – 50MΩ | Good | Acceptable |
| 50 – 100MΩ | Very Good | Acceptable |
| 100MΩ+ | Excellent | Acceptable |

# Detailed Inspection
## Is the Type of Cable Appropriate?

Code B1

Introduction:

When we ask if the type of cable is appropriate we have to look at several points e.g. what is the sheath made of, what type of cable protection, is the cable static or flexible etc. For instance, whether the cable is static or flexible might have a bearing on what type of gland is going to be used. What the sheath is made of could have a bearing on what type of chemicals will attack it if used on a chemical complex. Let us look at several different cables and how they might be used:

Flexible Cable (or Flex):

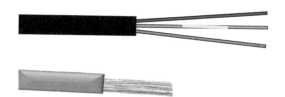

Exactly what it says, the cable is very flexible and can be used on portable appliances like hand-lamps, kettles etc. where the cable does not require protection like steel wire armour (SWA) or braid.

Another heat resistant flex is butyl rubber. Manufacturers of bulkhead light fittings usually use this type of flex to feed the lamp-holder from the connection block. The sheath is usually made of polyvinyl chloride (PVC) or cross-linked polyethylene (XLPE). The actual cable is very multi stranded to make it flexible. The gland for this type of cable would be a **compression gland** (or stuffing gland as technicians refer to it) as there is no braid or SWA to earth. If this cable is installed on site then it must be protected by some means such as trunking, conduit etc.

Steel Wire Armour Cable (SWA):

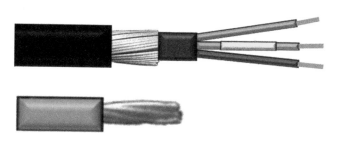

This cable is a bit more substantial and has the SWA as a cable Protector. Of course the SWA requires earthing and that is done by using a universal gland for SWA/braid. Not as flexible as the one above.

This SWA cable is usually round **extruded bedded** which means the bedding (the inner PVC) is manufactured in one continuous length so it cocoons the conductors the full length of the cable. The sheath is usually polyvinyl chloride (PVC) or polyethylene (PE). The polyethylene is much harder and can withstand many chemicals which would affect the PVC. Cross-linked polyethylene (XLPE) surrounds the conductors. The bedding (inner insulation) is Polyvinyl chloride (PVC). The cable can be obtained with aluminium wire armour (AWA) instead of steel wire armour (SWA). The steel wire armour should be earthed but not used as a sole earth (company policy).

# Detailed Inspection
## Is the Type of Cable Appropriate?

### Code B1-1

### Braided Cable:

Braided cable has gradually become more popular than steel wire armour cable for wiring fixed systems in industry. The sheath can be ethylene propylene rubber (EPR).

There are two types of braided cable, the one used for fixed wiring systems similar to SWA and the braided cable used for temporary supplies which is much more flexible. If the question was asked 'Why Braided Cable and not Steel Wire Armoured Cable' the answer might well be 'Braided Cable is lighter and more flexible'.

### Mineral Insulated Copper Covered (MICC):

MICC cable is a copper tube with conductors suspended in a white powder called magnesium oxide. Back in the 1960s and 1970s this was the only true flameproof cable as the magnesium oxide was the perfect filler since no explosion could get past it, therefore chemical plants tended to use it.

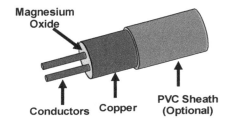

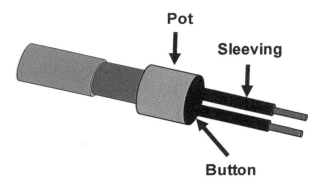

The magnesium oxide is very hygroscopic (water absorbent) so what is called a pot has to be put on the end filled with compound to waterproof the cable end.

The compound and the way the pot is made off differs between an Exd gland and an Exe gland.

Back in the past I (the author) have made hundreds of MICC ends off and we found that the glands for this cable are Exd even though the pot may be made off Exe. See the MICC instruction manual for this type of cable.

# Visual Inspection
## Any Obvious Damage to Cables?

Code B2

Introduction:

When I mention here 'damage to cables' on a **Visual** Inspection, the Inspectors would need to know in their remit just how far up the line do they go? Inspectors cannot be expected to examine every metre of cable in great detail on, say, a 500 metre run, so a certain length from the equipment will be inspected in detail. The inspector may walk the entire run to see if there is any conspicuous catastrophic damage.

## Is there any local damage to Power Cables?

The cable trauma is, thankfully, not usually as devastating as in the diagram above. Damaged power cables can be catastrophic to the plant as you can imagine. If the cable was damaged whilst isolated and the electrical technician replaces the fuses in, say, a lighting, motor or socket circuit then there would be a cloud of sparks where the damage was, making it imperative that any **Visual** Inspection makes every effort to pick this up.

Inspectors need to check to see if they can spot any local damage to the cables, cable tray or whatever cable management is being used e.g. ladder rack, conduit, trunking etc. They may find damage with steel wire armour showing, or even conductors in extreme cases as above.

### ON FINDING SOMETHING LIKE THIS DO NOT TOUCH IT.

If any catastrophic cable damage is found, it must be reported to the control room and engineer **IMMEDIATELY**, not just noted in some documentation. The circuit must be isolated as soon as possible, but in a correct, controlled manner. If you were to just pull the fuses you may send the plant into a crash shutdown, creating many other problems. These days the plant is most likely a Zone 2 area so not expecting gas to be present. If the cable is showing steel wire armour (SWA) on a vertical run it must be reported straight away as water can enter here and bypass the outer sheath seal on the gland.

# Visual Inspection
## Any Obvious Damage to Cables?

## Code B2-1

### Is there any local damage to IS Cables?

The cable trauma is thankfully not usually as devastating as in the diagram above. Intrinsically safe cables (Light Blue) showing damage should again be reported immediately. Although probably not dangerous as far as sparking and ignition is concerned, this type of damage could cause instruments to malfunction, the plant to go into shutdown etc. Can Inspectors see any damage to the cables, cable tray or whatever cable management is being used e.g. ladder rack, conduit, trunking etc? They may find damage with steel wire armour showing or even conductors in extreme cases as above. On finding something like this **do not touch it**. If any catastrophic cable damage is found, it must be reported to the control room and engineer **IMMEDIATELY,** not just noted down in some documentation.

### Is there any damage to cable at the Gland?

The situation on the diagram to the right might be more common where the steel wire armour is showing at the gland. This might be because someone has not made the gland off properly in the first place. If the cable is showing steel wire armour (SWA) or braid on a vertical run it must be reported straight away as water can enter here and bypass the outer sheath seal on the gland.

### Example of Water entering Cable Sheath?

I came across a situation where one of our motors was filling with water in its terminal block and no-one could find out where the water was coming from. The cable had a polyethylene sheath, which is extremely hard and where it came off the cable ladder rack many feet above the motor there was a huge split in the sheath as in the diagram on the left.

So water was entering here and travelling many feet down to the motor, bypassing the gland and into the terminal box. We taped up the flaw very thoroughly with special tape, solving both the mystery and the fault. I thought I would mention this although on a **Visual** Inspection there would be no access equipment.

# Visual Inspection
## Is the Sealing of Trunking, Ducts, Pipes & Conduits Satisfactory?

## Code B3

### Introduction:

Ensure that ducts & pipes are sealed. The sealing of the trunking/duct does not in fact come under Atex as they are classed as what is called **'Non-Sparking'**.

### Sealing Trunking/Ducts in Hazardous Areas:

The cables come through the duct as shown left, and a manufacturer's resin is injected around the cable to make a seal as per instructions.

So the company to whom the ducting belongs would end up with a **'Statement of Exclusion'**.

What the seal is achieving is stopping the migration of water and/or gas into the switch room or substation. The sealant can be fire proof and also act as a fire barrier as well as being vermin proof.

### Stopper boxes for conduits:

Ensure that stopper boxes are fitted in conduits where appropriate. Stopper boxes may not look as they sound. They may be 'inline' units as below.

Stopper boxes are compound filled 'tubes' that fit into the conduit system as in the picture, right. These are filled with manufacturer's compound and designed to stop explosions passing along the conduit. The 'box' can be opened for inspection of the compound. On a **Visual** Inspection the Inspector can only see if they are present.

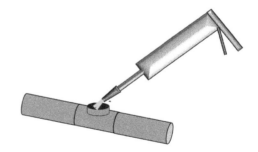

In my experience, conduit systems are not common in hazardous areas these days, but if they are used then there should be a stopper box in the conduit as it leaves the hazardous area, to stop explosions passing into a non-hazardous area like a switch room.

# Detailed Inspection
## Are all Stopper Boxes and Cable Boxes Correctly Filled?

## Code B4

### Introduction:

Electrical isolation required and Gas Free Certificate. Stopper boxes serve two purposes: they stop explosions travelling down the conduit, but they also restrict the passage of gases and vapours from travelling down the conduit.

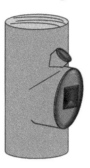

They can be in a conduit system right on the edge of the hazardous area. This stopper box is to prevent any explosion which may be travelling along the conduit from entering the non-hazardous area. They can be used when two items of flameproof equipment are joined together by conduit to stop the explosion from one box going into the other. They can also be used when a flameproof enclosure is joined by conduit to an enclosure that is not flameproof i.e. Exe, so that if there was an explosion in the Exd enclosure it does not travel into the Exe enclosure causing a catastrophic failure.

### Compound:

The large plug is removed from the middle of the fitting (above). If the fitting is vertical, a dam is put inside at the bottom to stop the compound from leaking out when it is poured in. Compounds have different properties depending upon the make. It is possible for instance to obtain a compound that sets in minutes, these are common. Some compounds can be used even if the temperature around the stopper box is extremely low.

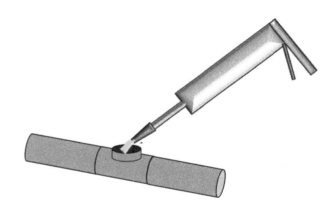

### Flexible Conduit Stopper Box:

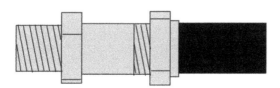

It is of course possible to get a flexible conduit stopper box, Exd or Exe, produced to Atex and the IEC60079 Standard.

# Detailed Inspection
## Is the Integrity of Conduit Systems in a Mixed System Maintained?

Code B5

Introduction:

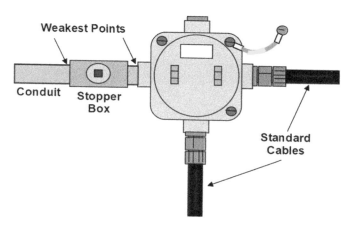

Looking at the above diagram where standard cables mix with conduit, **in my opinion** the following points should be followed & inspected:

1   The conduit should be made off into a **METAL** junction box. In this way it is easier to maintain earth continuity. (An Exd flameproof box would be ideal.)

2   A good idea might be to put a stopper box just before the junction box so that any explosion travelling down the conduit does not go through the cable glands and into the cable (especially if the conduit run passes through a Zone 1 area). Ensure that the stopper box is full of compound.

3   If the junction box is in a Zone 1 area and is made of metal it would presumably be an Exd flameproof junction box. The **5 full thread rule** must be applied to the threaded entry of the conduit into the junction box.

4   The thread of a conduit is always the weakest point where it enters into an enclosure so check for cracking.

5   Inside the junction box check that the cores are not 'scuffed' as they come out of the sharp conduit. (Ideally a sleeve should have been slid over the cores as far as possible.)

# Visual Inspection
## Is the Earth Bond Connected & is the Earth Wire Sufficient Cross Sectional Area?

## Code B6

### Introduction:

What is the difference between earthing and bonding? The answer may be, not a great deal. Equipment is earthed to give a path back to the star point of the electrical distribution transformer so that protective devices such as fuses or earth leakage circuit breakers can be operated in the event of a fault in the equipment which may otherwise leave it live. This is usually achieved by a third green & yellow core in the feeder cable. Bonding is mainly to stop two parts of a metal system from becoming different in potential (equipment, say, and the metal cable tray, and is usually achieved by an external green & yellow wire).

### Is the Earth Bond present on 'End of Line' Equipment?

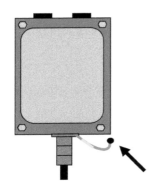

If the equipment is 'end of line', meaning that the circuit stops here such as in a single light fitting, technicians may fit an earth tag connected to earth. (Manufacturers may give advice. Check company policy here.) As you can see in the diagram (left) of the Exe increased safety or ExnR restricted breathing bulkhead light fitting, the earth bond is via an earth tag on the gland. It is unlikely that any bulkhead fitting other than Exd flameproof has a purpose fixing screw for an earth bond.

### Is the size of the Earth Bonding Conductors 4mm?

The earth bond wire shall be minimum of 4mm².

This cable must be capable of carrying any earth currents and at the same time offering least resistance.

### Has the Exe, Exn or Exi Equipment got a 4mm Bond to Earth?

If the equipment is Exe, Exn, Exi or uncertified then these items could be made of plastic. As you cannot use a bonding screw on plastic you have to achieve this another way by putting a brass earth tag on the gland. This would ensure a connection to earth of the equipment case, the cable SWA and the gland.

# Visual Inspection
## Is the Earth Bond connected & is the Earth Wire Sufficient Cross Sectional Area?

## Code B6-1

### Has all Exd Flameproof Equipment got a 4mm bond to Earth?

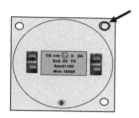

If the equipment does not have a direct bond to earth then you are relying on two things to achieve the task: one being the gland on the feeder cable, the other being the fixing screws, if on, say, the cable tray, that fasten the equipment in place.

If the equipment is Exd flameproof then there is usually an earth bonding screw to connect an earth bond to, and the other to a good earth as in the JB left.

### If the Equipment has Internal Earth Plate one 4mm Bond to Earth:

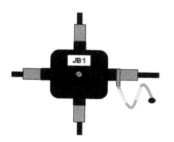

If the equipment, for example a junction box, has an internal earth plate (joining all of the glands together) then only one of the glands requires an earth tag & bond to earth as shown in the junction box left.

When you carry out the **Detailed** Inspection, **EACH** gland would have a serrated washer and locknut connecting it to the internal earth plate.

### If there is no Internal Earth Plate 4mm Bond on EACH GLAND:

If the Equipment, for example a junction box, has no internal earth plate as shown right, then an earth tag & bond must be put on **EVERY** gland.

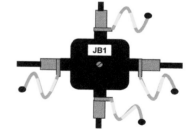

1   This can be done by adding individual earth tags & bonding each gland to earth or

2   Looping all tags together and taking one to earth.

I prefer number 1 because if the single bonding wire (2) develops a bad connection then none of the cables will be bonded. Check company policy.

# Close/Detailed Inspection
## Are All of the Earth Bond Connections Tight?

### Code B6-2

### Introduction:

On a **Visual** Inspection Inspectors are checking to see that there are earth bonds where these are supposed to be present. On a **Close/Detailed** Inspection Inspectors are checking to see that those earth bond screws are tight and there is a good connection.

On the IS barrier box the cover would not be removed on a **Close** Inspection so these connections, together with earths **INSIDE** power equipment, can only be checked on a **Detailed** Inspection.

### Slack Connections:

As above, on a **Visual** Inspection, Inspectors are checking to see if there is an earth bond present and whether it goes to a known earth. On a **Close/Detailed** Inspection, Inspectors can now get hold of those connections i.e. the one on the earth clip and the one on the known earth, say, the cable tray, and see if those connections are tight. Slack connections would mean a high resistance earth that may not do the job it is intended to do.

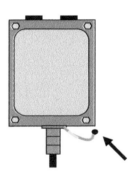

Also, in the case of a fault to earth with a slack connection, you could get sparking which is the last thing you want in your zoned area.

### Flameproof Equipment Bonding:

In the case of Exd flameproof equipment there is usually a dedicated earth bonding screw.

Both the bonding screw on the equipment and the screw on the cable tray, if that is being used as the earth, must be checked for tightness.

# Detailed Inspection
## Earth Loop Impedance of a TN System?
## (Terre Neutral Separate)

Code B7

Introduction:

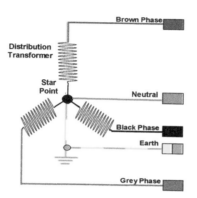

Why do an Earth Loop Impedance Test?

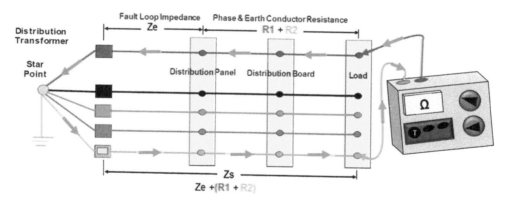

What are Inspectors doing with an Earth Loop Impedance Test? Well you could say that by utilising the live and earth conductor we are driving electrons through the star point of the distribution transformer as shown by the red arrows on the above diagram, and obtaining a reading in Ω of that circuit. Quite a high current is involved, around 20 amps. Tests are from the furthest point from the distribution transformer if possible.

We have to ensure that if a fault did occur in an installation, the current that flows would be enough to operate the protection systems such as circuit breakers or fuses and it must be ensured that these devices will operate in a certain time. Just imagine what could happen to hazardous area equipment if the protection devices did not operate fast enough:- **LIVE EQUIPMENT/HEAT/SPARKING.**

**Readings as per 18th Edition IEE Regulations (BS 7671).**

# Detailed Inspection
## Earth Resistance of an IT System? (Isolated Terre)
## Earth Resistance of TT System? (Terre Terre)

## Code B7-1

### Introduction:

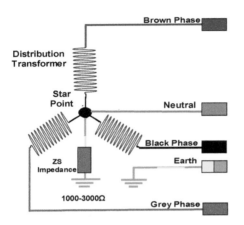

Here in the IT system a high impedance (Zs) is inserted into the distribution transformer star point earth. Sometimes there is no impedance at all so that the star point is not earthed.

The idea is that in the case of an earth fault on site, it would cut down the high amperage fault currents if the star point had either no earth or a high impedance one. TT systems (Terre, Terre) can be tested in the same way.

### Testing the Earth Rod:

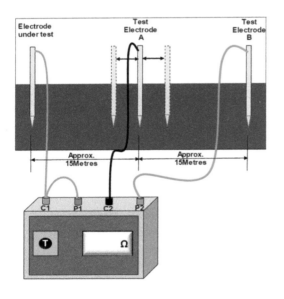

In the diagram to the left you can see the earth rod under test on the left hand side. Two test rods are then put into the ground, namely rod A & rod B. Rod B is put into the ground 25 – 50 metres away from the rod under test. Rod A is put into the ground about midway between rod B and the rod under test.

These rods are connected to the earth tester (Megger) as shown in the diagram, and a reading taken in Ω. Rod A is moved both nearer and further away from the rod under test as shown in the diagram, and at each position a reading is taken.

The most constant reading is the rod resistance that we are after. Many years ago in my day as a young electrical technician, this would be done with a bridge Megger as in Wheatstone bridge. The principle here is a current & potential loop, hence the C & P on the tester terminals, explained in my book 'Hazardous Areas for Technicians'.

# Detailed Inspection
## Automatic Electrical Protective Devices are Set Correctly?

Code B8

Introduction:

When it comes to automatic protective devices, electric motor overload units spring immediately to mind. These units can be set to, in most cases, **manual reset** and in other cases, **automatic reset**. The time delay to trip can be adjusted within a small band depending upon the full load current of the motor that they are protecting. A set of overloads are chosen to ensure that the full load current of the motor is around the middle i.e. if the motor full load current is 4 amps, a set of thermal elements might range from 2 amps to 6 amps.

NO Automatic Reset:

On a normal three phase, squirrel cage, induction motor the overloads would be set to **'Manual Reset'** meaning that, should the motor go into overload, the unit will trip, a small button will pop out and this must be then pushed in by hand to reset it. In the past if the motor was in a sump or well it might be that, for this purpose, the manufacturers would build an **Automatic Reset** into the overload unit. The motor could have thermistor units in the windings in case the motor got too hot.

**Automatic Reset** is, in my opinion, a very dangerous facility as if the motor did have any real problem with overload, the overload unit would just keep on resetting after a cool down period, and if the motor was also set to Automatic Start it would just keep starting and heat up. These days, especially on **Exe motors, the overload unit must NOT be set to Automatic Reset.** Automatic Reset may be a good idea on equipment such as fridges.

Current Transformer Overloads:

If the motor is quite large to save very large currents passing through the overload unit hence increasing its physical size enormously, it is possible for the manufacturers to build starters where the large currents pass through current transformers which in turn allows for physically smaller overload units.

Injecting using the CTs is called **Primary Injection** and requires a much larger injection set to obtain the current required in the CT primary ratio. **Secondary Injection** is preferred because of the smaller current injection.

**See next section B9 for connections and example injection currents.**

# Detailed Inspection
## Do Automatic Protective Devices Operate Within Permitted Limits?

## Code B9

### Settings:

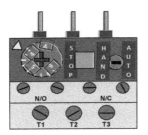

Overloads are set to the full load current of the motor obtained from the motor data plate.

This should tie in with the red load line or, as it is nicknamed, the 'Bloodline' on the ammeter.

### Secondary Injecting the Overloads:

Inject the overloads as per manufacturer's instructions which should be something like the following:

Connect all of the overloads in series as per the diagram on the right ensuring that the current flows in the same direction through each phase overload (or see manufacturer's documentation). Suggested injections: From cold, inject **150%** of full load current and see how long they take to trip (minutes). From hot, as soon as the overload will reset, inject **300%** of full load current (seconds). As soon as the overloads will reset, inject **108%** of full load current (should not trip in 8 minutes).

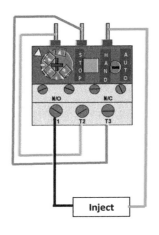

### Single Phasing Injection:

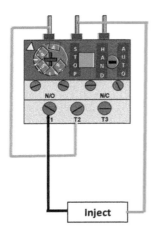

Inject the overloads as per manufacturer's instructions, which should be something like the following:

Disconnect one phase of the overload unit as per the diagram on the left (ie: the purple cable) and connect the Injection set as per the red and black cables. Now inject **105%** of the full load current and ensure that the overloads trip in a manufacturers' given time. This test is to ensure that the motor is protected from single phasing because, say, of one fuse blowing, because, of course if this happened, the current would rise on the other phases.

# Detailed Inspection
## Are All Conditions of Use Complied With?
## (Items with an 'X' after the Atex/BASEEFA Number)

## Code B10

### Introduction:

This section refers to a scenario when the equipment has an **'X'** after the Atex number. The **'X'** refers to **'Special Conditions of Use',** which will also be stated on the Certificate which comes with the equipment. The Inspector must look at the Certificate, note the Special Conditions, and see that they are being complied with.

### Where is the 'X'?

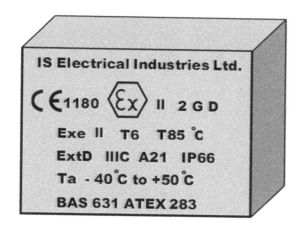

As you can see on the diagram to the left, the 'X' follows the Atex or BASEEFA Number.

As above, what you are looking for now is the Certification that came with the equipment so that you can see what the Special Conditions of Use are, and then see that they are being complied with.

It is very important that the installer is shown the Certification so that they are aware of what Special Conditions they have to abide by when installing the equipment.

### Three examples of Special Conditions:

| | | |
|---|---|---|
| Enclosures assigned an impact level of 4J or 4NM and must be installed only in areas of low mechanical danger | Precautions must be taken to ensure that the thickness of dust layer on the terminal box will not exceed 5mm | Wiring within the enclosure must not be grouped or bunched to prevent hot spots forming |

# Detailed Inspection
## Are All Cables Not In Use Terminated Correctly?

## Code B11

### Introduction:

If an item of equipment is removed for any reason e.g. an electric motor or a light fitting, then the cable must be correctly isolated from the supply and insulated or earthed, or failing that, terminated in a correct protection enclosure. It is really up to company policy and the electrical engineer in charge.

### Lighting Circuit 1:

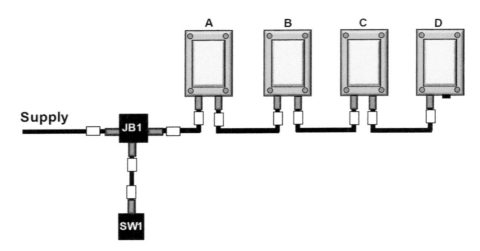

Looking at the lighting circuit above, let us say that we want to disconnect and remove **light 'D'**. It would be fairly difficult to do this. Remember the cables between fittings could be quite a few metres long and run on cable tray. This is not the usual way to connect lighting in a hazardous area. Why? Let us look at the following three scenarios of disconnecting one light.

1   We could isolate the circuit, remove the wiring from light 'D' and leave the cable connected to light 'C'. Take down the fitting. Tape the ends of the cable and put it in a plastic bag. In this case **we could not put the power back on** to lights A, B & C.

2   We could isolate the circuit, remove the wiring from light 'D' and leave the cable connected to light 'C'. Take down the fitting, put a correct certified junction box in place of light 'D' with the correct stoppers and put the cable ends into the connection block. In this case **we could put the power back on** to lights A, B & C.

3   We could isolate the circuit, remove the wiring from light 'D', disconnect and remove the cable from light 'C'. Take down the fitting, tape the bare ends of the disconnected cable and put a taped up plastic bag on either end. Put a correct certified stopper in light 'C' where the cable was removed. In this case **we could put the power back on** to lights A, B & C.

# Detailed Inspection
## Are All Cables Not In Use Terminated Correctly?

## Code B11-1

### Lighting Circuit 2:

Now on this page we can look at another way that the lighting can be connected making it much easier to remove a fitting. Simply go into JB 1, 2 or 3 disconnect the cables and tape them inside the JB, replace the lid and remove from the fitting.

Either tape the ends and plastic bag, or fit a correct certified JB with correct glands and stoppers whichever is company policy.

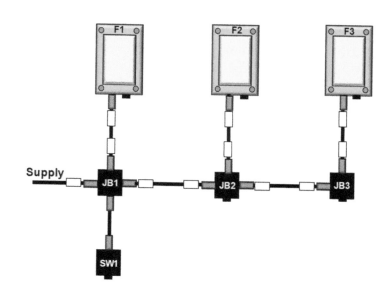

### Electric Motor Disconnection 1:

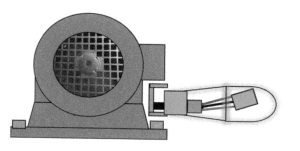

The electric motor to the left has been disconnected with the cable ends taped and a plastic bag over the whole. One point here is that if the fuses were replaced by mistake and the start button operated, there could be a large explosion inside the plastic bag.

### Electric Motor Disconnection 2:

The electric motor to the right has been disconnected with the cable connected into a correct Certified junction box. One point here is that if the fuses were replaced by mistake and the start button operated, there should not be too much sparking at the motor end.

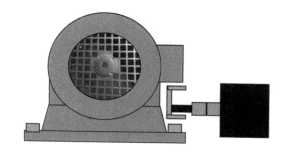

# Visual Inspection
## Any Obstructions Close to the Flameproof Flange?

### Introduction:

If the correct distance is not left clear there is a risk of back pressure. This takes away one of the three protective actions of an Exd flameproof enclosure i.e. in an internal gas explosion: pressure going from a **HIGH PRESSURE** explosion to a **LOW PRESSURE** atmosphere.

The distances are:    **40mm for llC,**    **30mm for llB,**    **10mm for llA.**

Of course if a llC enclosure was to be used in a llA area then the distance would be the lower one of **10mm.**

### In the Web of a Girder:

Causing pressure piling by installing Exd equipment too close to an obstacle without leaving the correct clearance. Installing a flameproof junction box in the web of a girder as shown in the diagram on the right may not leave enough clearance on two sides of the box. The correct clearances should be left on **ALL SIDES** of the equipment.

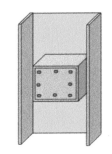

I did hear of one company that installed several Exd flameproof stop units in the web of a girder, and the Factory Inspector made them take the units out.

### Up against a Wall:

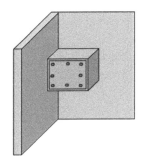

Causing pressure piling by installing Exd equipment too close to an obstacle without leaving the correct clearance. This is easily done if you are short of space or the enclosure protrudes significantly. Installing a flameproof junction box close up to a wall as in the diagram on the left may not leave enough clearance on three sides of the box. The correct clearances should be left on **ALL SIDES** of the equipment.

### Exd Flameproof Enclosures in a Line too close together:

I have seen the example (right) done with a row of stop/start buttons for a line of motors. The middle enclosure did not have enough clearance on **ALL** sides. **40mm is very roughly around 1.5 inches just to give you an idea.**

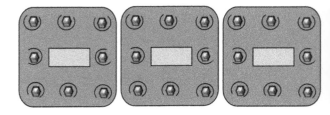

# Close/Detailed Inspection
## Variable Voltage/Frequency Installation in Accordance with Documentation?

### Code B13

### Introduction:

To vary the voltage in AC usually requires the involvement of some kind of transformer, for instance an auto-transformer. In the past, in an AC electric motor sometimes speed control was achieved by this varying of the voltage. The problem here of course is torque, decrease the voltage at the motor terminal block too much and there will not be enough torque to drive whatever the load is. So these days varying the voltage and the frequency by inverter is preferable.

So variable frequency (VFD) or variable voltage/variable frequency (VVVF) is what we are discussing. For this, the AC is taken into a converter, changed to a stable DC to smooth it and then into an inverter and this is where we get the variable voltage/frequency. As you can imagine, it is not a cheap set up. The motors range from low voltage (415 volts) up to high voltage (3.3Kv)

### Maintenance:

As you might imagine, maintenance is very limited if the drive is high voltage. So low voltage drives (415 volts) are what we will discuss. The system that we have here changes the power to the motor stator. We will call the converter & inverter combined the **'Unit'**.

What do we look for on an inspection of the **'Unit'**? Well firstly let us look at the four golden rules that I deem most important and then discuss them. Firstly: **CLEANLINESS**, keep the vital components clean, secondly: **DRYNESS**, keep the air around the components dry, thirdly: **TIGHTNESS** of all screw connections, and finally: **KNOWLEDGE** of the system.

### What Certification is the 'Unit'?

If the **'Unit'** is Exd or Exe there is very little inspection that can be done. For instance you can see if there is a build up of dust or dirt, are there any bolts that are slack in any of the covers, does the unit look corroded, is a coat of paint required, is there any damage that is outstanding, any corrosion?

If the **'Unit'** is in a building then the maintenance Inspection tasks would go up slightly. Again, this is a **Close** Inspection so no isolation.

# Close/Detailed Inspection
## Variable Voltage/Frequency Installation in Accordance with Documentation?

### Code B13-1

### 1    Cleanliness:

If the system is Exd or Exe, on a **Close** Inspection the maintenance Inspection is very limited as you cannot open up the equipment. So you are just looking for a build up of dust/dirt on the equipment. If the **'Unit'** is in a separate room then the Inspection is much more intense. Remember we are on a **Close** Inspection. One of the maintenance issues that is detrimental to the hardware of the system is dust. Dust can smother electronics causing heat and cut down the flow of air over hardware such as heatsinks which take heat away from components that are doing a lot of work. Fans can get their airflow cut down, you only have to look at your PC Fan and see how dust & dirt collects on it. Dust on the outside of the **'Unit'** can be sucked inside, so report any build-up on the case of the equipment.

On a **Detailed** Inspection there are two ways to expel the dust, either by vacuum or by blowing air. I have always believed that by blowing air onto the dust the problem is just transferred to other areas. In addition, the air supply would have to be an Instrument Air supply that was perfectly clean and dry, also blowing the air along a surface may create static. In my opinion, from experience, vacuuming, although static charge may be created, is less harmful. You can buy anti-static sprays these days, but I would check with the manufacturer to see if this was suitable.

### 2    Dryness:

Again, if the 'Unit' is Exd or Exe, all you can do is check that it is protected from the weather as far as possible. Are all of the bolts present and tight? If the **'Unit'** is in a separate room then again your Inspection would be much more intense. Initially, the building where the **'Unit'** is located is very important. It should be located in an area where the air is fairly dry and unlikely to be contaminated by moisture or dust. The **'Unit'** may have filters which must be inspected and, if this is a **Detailed** Inspection, changed as necessary. Look at the components and see if, in your opinion, moisture has been affecting them. Check the room, is it dry and fairly dust free?

# Close/Detailed Inspection

## Variable Voltage/Frequency Installation in Accordance with Documentation?

Code B13-2

### 3 Tightness:

If the **'Unit'** is Exd or Exe then all Inspectors are looking for is the tightness of the bolts in the covers. If any bolts are slack, further investigation may be warranted. If the **'Unit'** is in a room then the maintenance Inspection would be a bit more intense. This would be a **Close** Inspection so you are limited as to what you can do. On a **Detailed** Inspection screw tightness can be checked. Look at the components of the **'Unit'** very carefully and see if there are any signs of loose connections. Connections become loose in many cases through warmth causing expansion and contraction of the metals. Vibration can also have an effect. **Follow Manufacturer's Instructions at all times.**

**FOLLOW CORRECT ISOLATION PROCEDURE AND PERMIT SYSTEM.**

### 4 Knowledge:

This type of **'Unit'** requires that Inspectors have a certain standard of knowledge and are not given the inspection to do if they have never worked on this type of equipment before.

**A FULL MANUFACTURER'S MAINTENANCE MANUAL IS VITAL IF THE INSPECTOR IS TO GO INSIDE OF THE 'UNIT' FOR ANY REASON.**

# Detailed Inspection
## Do Motor Protection Devices Operate Within the tE Time:-
## Exe Motors Only.

## Code B23

### Introduction:

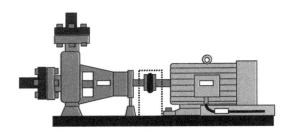

A typical pump unit is shown, left. The most common motor is a 3 phase, squirrel cage, induction motor, and the most common pump is a centrifugal pump.

Just imagine what would happen if the pump unit stalled for some reason and the motor was still connected to the supply? If it did not shear the rubber tyre coupling (sometimes called the 'Element') the motor would be connected and trying to spin.

### The tE Time:

The above situation is where the **tE time** comes into play and how the motor control unit (the starter) is armed to deal with it. Usually the tE time is around five seconds. The definition states that the tE time is the time taken, under locked rotor conditions, for the motor winding temperature to rise from its **Designed Maximum Running Temperature** to the **Maximum Allowable Temperature for the Class of Insulation used in its Construction.** The tE time can usually be found on the Exe motor data plate along with the letter denoting the insulation class. In theory, on an Exe motor, there are two separate times: one for the rotor and another for the stator, the shortest time of the two being taken as the tE time.

### The Overload Unit:

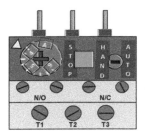

The overload is usually what protects the motor for the locked rotor tE time, although I have known the fuses to blow before the overload trip time. The tE time and motor start time must not be too close otherwise on repeated starts the overload could trip and there is no fault.

How to test this overload unit is shown in the 'Automatic Electrical Protective Devices' Code B8 & B9 earlier.

# Visual Inspection
## Is the Equipment Protected Against the Weather?

## Code C1

### Introduction:

We cannot put our hazardous area electrical and instrument equipment inside a bubble and protect it from all weathers. The equipment ingress protection (IP) will usually protect it against solids and liquid ingress. What about direct sunlight and, at the other end of the scale, extreme cold? Direct sunlight can obviously cause the equipment to heat up.

Older equipment has a temperature classification that usually requires the ambient temperature normal to be between **-20C to +40C.** More modern equipment has higher and lower ambient temperatures, but might be worth investigation. Another point to consider is that at very low and very high temperatures, certain electrical insulation materials will degrade and, to some extent, lose their insulating property. If the ambient temperatures are known to be higher or lower than the 'norm', equipment with Ta or Tamb can be obtained with a higher or lower ambient temperature.

### Very Low Temperatures:

The Inspector(s) should be concerned as to whether in very low temperatures the hazardous area equipment can maintain the 'Protection' that make it a Certified item e.g. Exd, Exe, Exi, by way of the material used to construct the equipment, the norm being -20C to +40C. So:

1   **Can the equipment function efficiently at very low temperatures, for instance if the temperature was to go below the norm of -20C?**

2   **Could the surrounding atmosphere gas/vapour change its properties at very low temperatures?**

3   **Can the equipment function efficiently at very high temperatures, for instance if the temperature was to go over the norm of +40C?**

### High Temperatures (e.g. Direct Sunlight):

Heat can cause electrical insulation degeneration. Cables in conduits which are in direct sunlight can be subjected to very high temperatures. PVC conduits & pipes can become more brittle when subjected to direct sunlight due to ultra violet (UV) light on the PVC. Heat can build up outside and inside enclosures, especially if they are made of metal. Manufacturers may give guidance on ambient temperatures.

# Visual Inspection
## Is the Equipment Protected Against Corrosion?

## Code C1-1

### Introduction:

Corrosion is usually caused by some kind of chemical reaction. In many cases it is because the equipment is the **'Anode'**. By using **'Cathodic Protection'** we can cause the equipment to be a **'Cathode'** instead. This can slow the corrosion down enormously. Below we talk about looking for corrosion during Inspections and what signs the Inspectors might be looking for.

### Inspecting Equipment for Corrosion:

Inspectors carry out the Inspection, if there is any corrosion they must Risk Assess:

1   **What is the corrosion and what equipment is it on?**

2   **Is the corrosion minimal and requires no action?**

3   **Is the corrosion acceptable at the time of Inspection but a Risk Assessment determines it will get worse in the near future?**

4   **Is the corrosion great at the time of the Inspection and requires maintenance work straight away?**

### What is Corrosion?

I always look at corrosion as being released energy that was trapped into, say, steel at its liquid state and the metal is now returning to a more stable state e.g. Oxides, Sulphides or Hydroxides. If we look at any rusty structure that is made of steel, for example a rusty pipeline without any cathodic protection, it is not all rust (Ferrous Oxide), there are some good areas here and there. Where those good and bad areas meet, a cell is formed between the good parts (Cathodes) and the corroding parts (Anodes), the electrolyte (Water) provided by moist air or soil.

### Cathodic Protection:

| Cathodic ↑ ↓ Anodic | Platinum |
|---|---|
| | Gold |
| | Titanium |
| | Silver |
| | Tin |
| | Lead |
| | Mild steel |
| | Aluminium |
| | Zinc |

We can use cathodic protection on a pipeline to slow this process down, explained in the book **'Hazardous Areas for Technicians' ISBN: 978-1-912014-95-8.**

If we take a cable tray, the galvanising is a type of zinc coating cathodic protection. If you look at the cathodic table left you will notice that zinc is more **'Anodic'** than mild steel hence the zinc forces the mild steel to be a **'Cathode'**. Corrosion will always go for the anode i.e. the zinc.

If you look at Gold you will see that it is near the top of the table i.e. very cathodic, which is why you can dig up gold coins out of the ground and there is hardly any corrosion at all.

# Visual inspection
## Is the Equipment Protected Against Vibration?

Code C1-2

Introduction:

There are many things that can cause vibration in chemical factories and on platforms. In the electrical world, electric machines spring straight to mind e.g. motors, generators etc. In the mechanical world there are compressors, pumps etc. In the process world there are the plant processes themselves.

Vibration effects on Electric Motors:

Let me ask a question: if I have two motors, an 'A' machine and a 'B' machine, if I put the **'A'** machine on **'RUN'** and the **'B'** machine on **'STANDBY'**, do I run the **'A'** machine to destruction knowing that I have a good **'STANDBY'**? The answer might depend on how much vibration there is on the **'STANDBY'** machine, and for how long it is going to be stopped.

If the **'STANDBY'** machine has been stopped for a very long period of time and there is a lot of vibration on the machine, you could end up with a phenomena called **'FALSE BRINELLING'** where the rotor acts as a hammer on the bearings. Lubricant (grease) is pushed out from between the ball/roller of the bearing so there is a metal to metal contact. With the rotor constantly hammering the bearing with the vibration, flats can be formed on the balls/rollers and indents pushed into the race. Hence when the motor eventually starts it will rattle itself to pieces and will not last very long at all.

On a **VISUAL** Inspection the Inspector can (if possible) only listen to the running machine. When it comes to the **CLOSE** Inspection there is a whole range of instruments to detect the level of vibration here, **but this is only motor, pump and compressor etc bearings.**

Vibration effects on Lighting:

Vibration can cause frequent failure in light fittings by shaking the filament of the lamp and causing premature failure. If the plant lights are going faulty quite regularly it might be worth checking the plant vibration levels and the effects on the lighting. Incandescent lamps can be vulnerable to vibration. Manufacturer's data may help.

# Visual Inspection
## Is There Accumulation of Dust or Dirt?

### Code C2

### Introduction:

If dust or dirt is allowed to build up on the equipment then this could have a detrimental effect on the temperature of the equipment. On a **Visual** Inspection you are only looking to see if there is a build-up of dust or dirt on the equipment.

### Temperature Classification in General:

| | |
|----|-----|
| T1 | 450 |
| T2 | 300 |
| T3 | 200 |
| T4 | 135 |
| T5 | 100 |
| T6 | 85 |

The table to the left shows the six temperature classes in the UK. Equipment should not go over these temperatures under NORMAL AND SPECIFIED FAULT CONDITIONS. The ambient temperature for these temperature classes has a norm of -20C to +40C

If the Ambient Temperature is higher or lower then alternative equipment must be obtained and the data plate will state Ta or Tamb in which case the new higher and/or lower ambient temperature will be stated.

### How Hot are these Temperatures?

These temperatures are taken for granted, but do you realise how hot the equipment would be if it reached 450C as in T1? Look at the effects that temperature has on certain materials, some of which are used in the electrical world.

In the table on the right you can get some idea of what can really happen if things start to heat up due to fault conditions.

| | |
|-----|--------------------|
| 455 | PVC Ignites |
| 327 | Lead Melts |
| 232 | Tin Melts |
| 135 | Polyethylene Melts |
| 100 | Water Boils |
| 85 | Burn Hands |

WE DO NOT WANT ANY CONDITIONS THAT MAY CAUSE THE EQUIPMENT TEMPERATURE TO RISE, AND EQUIPMENT COVERED IN DIRT AND DEBRIS IS ONE OF THEM.

# Visual Inspection
## Is There Accumulation of Dust or Dirt?

## Code C2-1

### Dust Ignition Temperatures:

If we just work on the golden rule that 'clouds explode and layers burn', the table (right) shows the ignition temperatures of various dusts with the different ignition temperatures of their clouds and layers. Our equipment temperature must not get anywhere near to these temperatures or there could be an explosion (Cloud) or fire (Layer). Excessive dust or dirt beyond the 5mm maximum could smother the equipment and cause it to heat up so good housekeeping is vital.

| Dust | Ignition Temperatures °C | |
| --- | --- | --- |
| | Cloud | Layer |
| Cellulose | 520 | 410 |
| Flour | 510 | 300 |
| Grain | 510 | 300 |
| Sugar Dust | 490 | 460 |
| Tea Dust | 490 | 340 |
| Starch | 460 | 435 |
| Lignite | 390 | 225 |

### Dust & Dirt on IS equipment:

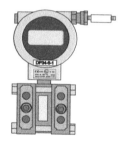

Dust and dirt on IS Instruments will not usually cause heat as there is no real power there. What it may cause, especially in the case of a transmitter, is for the instrument to malfunction. For example, sometimes an instrument takes its reference from atmospheric pressure. If any debris clogged up the reference part of the instrument it would malfunction. Dust and dirt on metal equipment can also cause corrosion so good housekeeping is essential.

### Dust & Dirt on Exp Pressurised Equipment:

It is extremely important not to get dust or dirt inside of pressurised equipment as it could contain sparking or hot components. The equipment and the immediate surroundings should be cleaned first to avoid any risk of dust entering the enclosure when opened for Inspection. After the Inspection the equipment should be cleaned internally to remove any remaining dust inside before the power is restored. This will be discussed in the Exp pressurisation section.

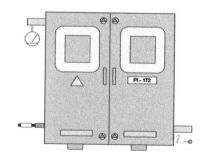

# Detailed Inspection
## Is the Electrical Insulation Clean & Dry?

## Code C3

### Introduction:

Electrical isolation required and Gas Free Certificate. Ensuring that the electrical insulation is clean and dry is vital for maintaining the integrity of the equipment. Damp insulation can cause tracking or explosions inside the enclosure. Flameproof items can suffer because of course flameproof enclosures are not waterproof although non-setting grease will go a long way to improve the IP. Exe and Exn enclosures will suffer with internal damp if the seals are not too good which is why it is important that these are checked. We do not want tracking and arcs and sparks in Exe or Exn enclosures.

### Flameproof Enclosure:

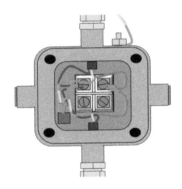

Now the equipment has been opened up the enclosure must be checked for moisture. This task is very important as it may be years before the equipment is opened again for another inspection or for any maintenance. As mentioned above, flameproof equipment is not waterproof so grease on the flameproof faces assists in the water sealing of the equipment. I have opened up a flameproof box before today and water has poured out, but I have also seen flameproof motors running under water in a pit.

### Exe/Exn Enclosures:

It is most important in Exe or Exn enclosures to exclude damp/water. Much trouble is taken here by manufacturers to ensure that there is no heat or explosions within these enclosures but dampness can cause just that. Usually, seals are the problem that will be found on Detailed Inspections. Check the lid seal.

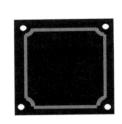
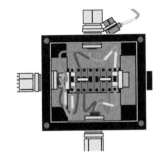

### Exi/Exe Enclosures:

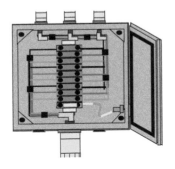

Although with Exi there will be no trouble with flashes and bangs, at least none strong enough to ignite a gas, vapour or dust, dampness can cause intrinsically safe circuits to simply not work or, even worse, give false readings. With an Exe box there could be a problem. Again with this type of enclosure it is usually a faulty seal that is causing the problem. Check the enclosure lid seal for damage or whether it is missing.

# Extra Points for Equipment Protection Exd, Exe, Exn & Ext:

The official IEC60079 Standard Codes go from **A1 – A31, B1 – B23** and **C1 – C3**. This is where the codes end.

I have studied the codes and have come to the conclusion that there are some points that may require an Inspection that are not in the codes. I have listed them below and the following pages describe my new points which I have called **E1 - E10**, **'E'** meaning **'Extra to the Standard Codes'**. You will see what I mean as you look though the list below:

E1     **(Visual)** Is there any damage to the outside of the equipment?

E2     **(Visual)** Is the **equipment** ID available? **(Tag, Plant Number etc.)**

E3     **(Close)** Is the **equipment** ID correct? **(Tag, Plant Number etc.)**

E4     **(Close)** Are the motor bearings satisfactory? **(Shock Pulse Test)**

(Company Policy if this Inspection is completed.)

E5     **(Detailed)** Does the inner insulation enter the enclosure?

E6     **(Detailed)** Are the gland internals correct when glands are split?

(Company Policy how many are checked.)

E7     **(Detailed)** Is the motor balance reading satisfactory?

(Company Policy if this Inspection is completed.)

E8     **(Detailed)** Is the motor earth path reading satisfactory?

(Company Policy if this Inspection is completed.)

E9     **(Detailed)** Is the switch wire marked?

E10    **(Detailed)** Is there heat resistant sleeving on the wires in a bulkhead light fitting? This has been removed from the Checklist as modern cables will withstand the heat. In many cases the insulation may be cross linked polyethylene (XLPE) which of course is heat resistant, low smoke and '0' halogen.

# IS Code Descriptions

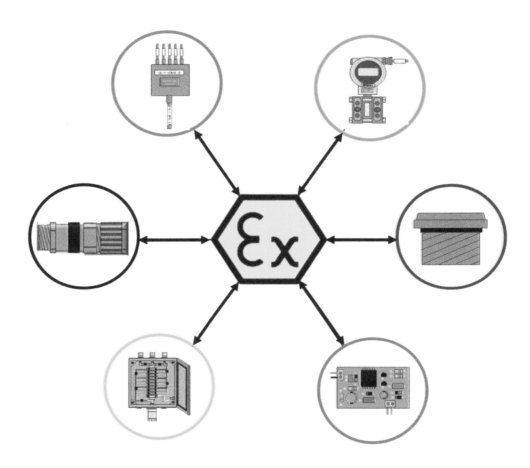

# Visual Inspection
## Is the Equipment Appropriate to the EPL/Zone Requirements of the Location?

## Code A1

### Introduction:

All that the Inspectors will have with them here is an **Area Classification Drawing.** This is a drawing with lines, circles or squares on it depicting the Zones. **No tools or access equipment.**

### Area Classification Drawing Markings:

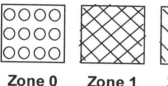

**Zone 0**    **Zone 1**    **Zone 2**

The **Gas Zone** markings will be shown on the **Area Classification Drawing** as shown to the left. These days Zone 2 will be the most common. Zone 0 usually inside of tops of vessels.

The **Dust Zone** markings will be shown on the **Area Classification Drawing** as shown to the right. Zone 20 usually inside tops of hoppers etc. Because of PPM Zone 21 areas will be rare.

**Zone 20**    **Zone 21**    **Zone 22**

On this drawing there will be no Gas or Dust Groups, no temperature classifications or surface temperatures, so although this information will be on the equipment there will be no reference on the drawing to match up.

### Protection/Equipment Protection Level (EPL):

What the Inspectors are looking for here is the 'Protection', i.e. Exia, Exib, Exic or ExnL on the equipment and to see if it is in the correct Zone **(see chart on next page).**

Manufacturers have started putting the EPL on the end of the markings, probably to get technicians and Inspectors used to seeing them. These are of course being introduced by the IEC probably with a view to replacing the category. Gas Zone EPLs are Ga, Gb or Gc. As can be seen from the above diagram this **EPL is 'Gb'** i.e. suitable for **Gas Zone 1.** Dust EPLs will be Da, Db or Dc.

It must be noted that equipment suitable for Dust Zones will not automatically be suitable for Gas Zones and vice-versa. So if, say, the Inspector was to find **EPL Gb** in a **Dust Zone 21** they should note this as a fault.

# Visual Inspection
## Is the equipment appropriate to the EPL/Zone requirements of the location?

## Code A1-1

| Protection: | Description: | Special Comments: | Zones: |
|---|---|---|---|
| Exd | Flameproof | Allows gas to enter | 1 & 2 |
| Exe | Increased safety | IP54 Minimum | 1 & 2 |
| Exec | Increased safety | Replaced **ExnA** | 2 ONLY |
| Exh | Mechanical | Combines Mechanical 'c' - 'b' - 'k' | Documentation |
| Exia | Intrinsic Safety | 2 Faults and remain safe. | 0, 1 & 2 |
| Exib | Intrinsic Safety | 1 Fault and remain safe. | 1 & 2 |
| Exic | Intrinsic Safety | Replaced **ExnL** | 2 ONLY |
| ExiD | Intrinsic Safety | Dust Atmospheres | Documentation |
| Exm | Encapsulated | Older type equipment | 1 & 2 |
| Exma | Encapsulated | Completely seals | 0, 1 & 2 |
| Exmb | Encapsulated | Completely seals | 1 & 2 |
| Exmc | Encapsulated | Completely seals | 2 ONLY |
| ExmD | Encapsulation | Dust Atmospheres | Documentation |
| ExN | Non Incendive | **Never Cenelec so no export.** | Withdrawn |
| ExnA | Reduced Risk | Non Sparking - Replaced by **Exec** | 2 ONLY |
| ExnC | Reduced Risk | Encapsulated | 2 ONLY |
| ExnC | Reduced Risk | Hermetically Sealed | 2 ONLY |
| ExnC | Reduced Risk | Enclosed Break | 2 ONLY |
| ExnR | Reduced Risk | Restricted Breathing | 2 ONLY |
| ExnL | Reduced Risk | Energy Limiting - Replaced by **Exic** | 2 ONLY |
| ExnZ | Reduced Risk | Pressurisation - Replaced by **Expz** | 2 ONLY |
| Exo | Oil Filled | Quenches any arcs or sparks | 1 & 2 |
| Exop (is) | Optical Radiation | Inherent Safety (is) | 0, 1 & 2 |
| Exop (sh) | Optical Radiation | Not Inherently Safe and Interlocked (sh) | Documentation |
| Exop (pr) | Optical Radiation | Protected Optical System (pr) | Documentation |
| Exq | Quartz / Powder Filled | Quenches any arcs or sparks | 1 & 2 |
| Expx | Pressurisation | Auto Shutdown | 1 & 2 |
| Expy | Pressurisation | No Shutdown | 1 & 2 |
| Expz | Pressurisation | No Shutdown - Replaced **ExnZ** | 2 ONLY |
| Expv | Pressurisation | Auto Shutdown | Documentation |
| ExpD | Pressurisation | Dust Atmospheres | 21 & 22 |
| Exs | Special Protection | **Older type equipment**. Never Cenelec | Withdrawn |
| Exsa | Special Protection | New Atex Certification | 0, 1 & 2 |
| Exsb | Special Protection | New Atex Certification | 1 & 2 |
| Exsc | Special Protection | New Atex Certification | 2 ONLY |
| ExtD | Protection by Enclosure | **Dust Atmospheres only** | Documentation |
| Exv | Ventilation | American Ex Standard | Documentation |

# Close Inspection
## Is the Equipment as Specified in the Documentation?

Code A2

Introduction:

Every item of equipment on a chemical factory or platform should have a number of some kind so it can be easily identified and any documentation or Certification will have a numerical reference in case the equipment were to malfunction. These numbers must match the plant documentation such as Loop Diagrams. A Loop Diagram has to be as near as possible to 100% correct all of the time. All instrumentation in a particular loop would be referenced on this drawing.

Instruments:

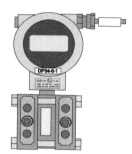

On the plant will be many instruments measuring level, flow, pressure, temperature etc. Each one of these instruments should have a tag number which identifies them on a Loop Diagram. This tag number will be unique so no other instrument on the plant will have the same number. On the **Visual** Inspection, the Inspector would just have an **Area Classification Drawing** and no Loop Diagram. With a **Close** Inspection Inspectors are also checking that the make, model and type numbers are correct as per loop diagram.

**ANY** differences from instrument to Loop Diagram are faults, either with the instruments or the Diagram.

Is the Box Number and Circuit Numbers Correct?

Marshalling boxes are large junction boxes out on the plant. There can be hundreds of different boxes with a whole range of different circuits going to the plant instrumentation e.g. IS Boxes, 24V DC Boxes etc. How they number these boxes is for plant/platform policy to decide. The number might include the box's own number which in the diagram on the right would be IS-1 (Intrinsically Safe Circuits) hence it must possess a Blue Label stating that the box only contains IS Circuits/Loops. The number might include the building where the main cable comes from/goes to, which in the diagram is Control Room 12 (CR12), the cabinet number where the main cable comes from/goes to in the diagram is number 2. All the loop cables must also have individual numbers. All Inspectors are interested in on a **Visual** Inspection is whether all numbers are present. On a **Close** Inspection they have a Schematic/Loop Diagram to check if the numbers are correct.

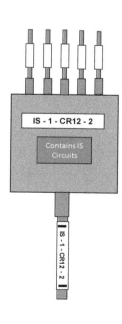

# Close Inspection
## Is the Equipment as Specified in the Documentation?

### Code A2-1

### Barrier Units:

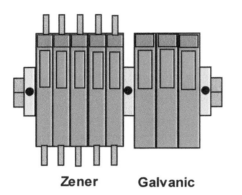

**Zener**     **Galvanic**

On the plant will be many instruments measuring level, flow, pressure, temperature etc. Each one of these instruments will have a barrier protecting it. This barrier make, model and type must match the numbers on the Loop Diagram.

With a **Close** Inspection, Inspectors are checking that these numbers are correct as per the Loop Diagram. **ANY** changes from instrument to Loop Diagram are faults.

### Loop Information:

The first thing is to obtain the Loop Diagram of the loop that is to be checked as below:

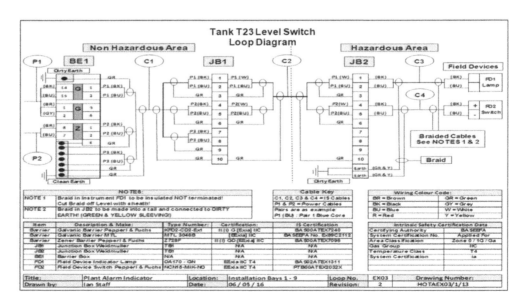

On the bottom of the Diagram should be the information on instruments, barrier units etc that are in the loop. If the site information differs from that in the Diagram, investigate which is correct. The instrument technician might have installed an uprated instrument and not changed the Diagram. So if, for instance, on site a barrier unit was 728F and on the Diagram it is shown as a 728E, then this would be incorrect and requires investigation (and advice as to remedial action).

# Close Inspection
## Is the Equipment Group Correct?

### Code A3

#### Introduction:

What the Inspector is looking for here is the **'Group'**. This can take several forms e.g. **Surface Industry** or Mining, the **Gas Group** and the **Dust Group**. If the equipment is Atex the Group will be in the form of Roman numerals after the Atex mark 'l' or 'll'. The predominant gases in each Gas Group are llA-Propane, llB-Ethylene & llC-Hydrogen. Dust Group equipment cannot, without the manufacturer's approval in writing, go into Gas Zones and vice-versa.

#### Mining Equipment:

If the equipment is for Mining, the **'Group'** is the **Roman 1** after the Atex mark as shown in the diagram to the left. Ma indicates the top section in Mining.

#### Surface Industry Gas/Vapour Equipment:

As you can see in the diagram to the right the **'Group'** is the **Roman ll** Surface Industry, after the Atex mark.

#### Gas Group ll:

The diagram on the left shows the **Gas Group** as just a **llC** The **Roman ll** after the **Atex** mark is for Surface Industry.

If the **Gas Group** is shown as just a **Roman ll** then the equipment will be suitable for any **Gas Group ll**. If the equipment is shown as **Gas Group llB + H2** then it is a **llB** piece of equipment, but is also suitable for **Hydrogen** (Not Acetylene or Carbon Di-Sulphide). Find out the Gas Groups of all chemicals, gases & vapours on plant and ensure equipment Gas Group is correct (which should be to the worst one**), llC being the worst.**

#### Surface Industry Dust Equipment:

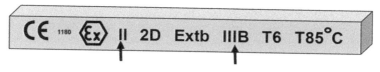

The **Group** will still be a **Roman ll** Surface Industry even though the Dust Groups are 'lll'. There is Dust Group lllA (Combustible flyings i.e. jute, hemp, oakum, kapok etc.), lllB (Non-conductive Dust e.g. wood, grain, starch, sugar etc.) or lllC (Conductive Dust e.g. coal, carbon, charcoal, magnesium etc), **lllC again being the worst.**

# Close Inspection
## Is the Equipment IP Rating Appropriate to the Group III Material Present?

### Code A4

### Introduction:

When we talk about Group III materials we are of course talking about Dust Groups IIIA, IIIB & IIIC and there are some examples of the actual dusts/flyings below. The 5X/6X refers to the IP protection which of course depends upon the Zone as well as the Dust Group. This could depend upon whether IP washers are required or not. However Dust Group IIIC is conductive dust, which is the worst case scenario and automatically becomes IP6X minimum.

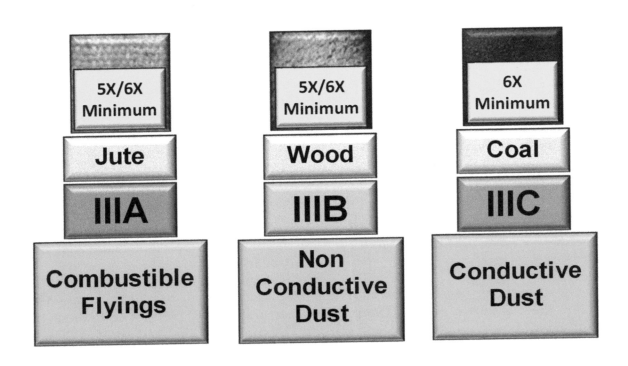

As pointed out in other chapters, in the dust world sometimes there is no second number in the IP code. The second number is replaced by an 'X' so instead of IP54 the equipment would be IP5X, therefore the Inspector would need to know what kind of dust is in any particular area and then check the equipment IP.

# Close Inspection
## Is the Equipment Temperature Classification Correct?

Code A5

Introduction:

This of course cannot be completed until the **Close** Inspection as the information required would not appear on the **Area Classification Drawing.** In this country we have six temperature classifications. The definition is: the surface of the equipment in contact with the gas, vapour or dust will not go over the temperature classification (T1 – T6) **under normal or specified fault conditions.** (It does not say it will never go over, and the manufacturers will specify the faults.)

Ignition Temperature:

| T1 | 450 |
|----|-----|
| T2 | 300 |
| T3 | 200 |
| T4 | 135 |
| T5 | 100 |
| T6 | 85 |

Find out the **Ignition Temperatures** of all of the chemicals, gases, vapours and dusts on the plant and ensure that the **Temperature Classifications** of all the items of electrical equipment are below that value. In reality, Electrical Engineers will order T5 or T6 equipment to be on the safe side, even though this may be more expensive.

For example if you had a gas or vapour that had an **Ignition Temperature** of 235C. Looking at the UK chart left it would be no good using equipment with **Temperature Classification** T2 (300C) because if the equipment was to reach 300C under normal or specified fault conditions and there was a gas cloud with a 235C **Ignition Temperature** the 300C equipment would ignite it.

The **Temperature Classification** of the equipment must be **under** the **Ignition Temperature** of the gases in which it is installed.

Ambient Temperature:

The above definition applies in a normal **Ambient Temperature of -20C to +40C.** So if the equipment does not specify anything else on the data plate then this is the **'Norm'.**

But what if the Ambient Temperature where the equipment is to be installed is higher or lower than the 'Norm'? This situation needs different equipment so it's back to the manufacturers to obtain this new equipment. The data plate will show the higher or lower **Ambient Temperature** requested with **'Ta'** or **'Tamb'** showing what the new higher or lower Ambient Temperature is, e.g. **Ta = -50C to +55C.**

# Close Inspection
## Is the Ambient and Service Temperature Range Correct for the Apparatus?

## Code A6 & A7

### Introduction:

The barriers will have an ambient temperature range. Remember these barrier interface units are protecting the instruments out on site and being 'associated' equipment they will not have a direct temperature classification, but in the documentation it should recommend an ideal ambient temperature. On the **Close** Inspection the documentation can be checked.

### Ambient Temperature:

Unless the documentation was to say different the ambient temperature for this equipment is as for most equipment, **-20C to +40C** and we can call this the **'Norm'**. If the ambient temperature is higher or lower than the **'Norm'** then users must go back to the manufacturer for some different equipment that will fit into the higher or lower ambient temperature. This new equipment will have on the data plate **'Ta'** or **'Tamb'** and what the new ambient temperature range is e.g. **Ta = -40C to +55C.**

### Simple Apparatus:

Simple apparatus should be non-heat generating equipment such as switches, junction boxes, LEDs, thermocouples & photocells provided they do not generate or store more than 1.5 Volts (V), 100milli Amps (mA), 25milli Watts (mW) & 20 Micro Joules ($\mu$J). Again the equipment may not have a temperature classification, but the manufacturers may suggest an ambient temperature.

### Service Temperature:

If equipment has a service temperature, i.e. the steady temperature when the equipment is operating at full capacity, then this temperature should be in the manufacturer's documentation. This temperature can be checked with a temperature gun when the equipment is operating.

### Storage Temperature:

The **storage temperature** for interface units such as Zener barriers is: **-25C to +70C.** Note: This is while units are on the Stores shelf and not being used.

# Close Inspection
## Is the Service Temperature Range Correct for the Apparatus?

### Code A6 & A7-1

#### Boilers:

In industry there are processes that definitely will produce heat e.g. industrial large boilers, reformers, kilns etc. You might think that these are not hazardous areas, however in some cases they might be right in the middle of one. Remember they also require gas (Methane) to work. These structures have instrumentation measuring a range of things e.g. temperature, gas flow, gas pressure etc. The ambient temperature may be well above the 'Norm' so correct instrument selection is vital.

#### How does extreme temperature affect apparatus?
#### Pressure Gauges:

Expansion and sweating inside instruments such as pressure gauges can cause metal joints to loosen, causing the gauge to give false readings and eventually the joint will fail completely. Always ensure that the service temperature is known prior to selecting and installing instruments.

#### Printed Circuits:

If printed circuit boards are heated it could cause the board to warp and damage the very delicate soldered joints and components. Sometimes faults on these boards are very difficult to locate. Advice on operating temperatures can be obtained from Manufacturers.

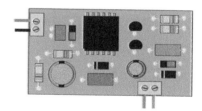

# Close Inspection
## Is the Installation Clearly Labelled?

## Code A8

### Introduction:

On the **Visual** inspection the Inspectors could only look to see if the equipment ID was present, they had no documentation with them to check that it was correct. All they had was an Area Classification Drawing. On the **Close** Inspection the Inspectors are checking that the equipment ID is correct as per the Loop Diagram.

### Is the Box Number and Circuit Numbers Correct?

Marshalling boxes are large junction boxes out on the plant. There can be hundreds of different boxes with a whole range of different circuits going to the plant instrumentation i.e. IS boxes, 24V DC boxes etc.

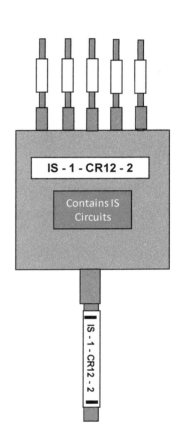

How the company numbers these boxes is for plant/platform policy to decide. The number might include the box's number which in the diagram on the right would be IS-1 (Intrinsically Safe Circuits) hence it must possess a blue label stating that the box only contains IS circuits/loops and no others.

The number might include the building where the main cable comes from/goes to, which in the diagram is Control Room 12 (CR12), the cabinet number where the main cable comes from/goes to in that Control Room in the diagram is number 2. All the loop cables must also have individual numbers.

So to sum up the box is IS-1, the main cable comes/goes to Control Room 12, cabinet 2.

As mentioned above all Inspectors were interested on a **Visual** Inspection was whether all **numbers** are **present**. On a **Close** Inspection they have a Schematic/Loop Diagram to check **if the numbers are correct**.

# Visual Inspection
## Is there any Damage to Glass to Metal Seals?

Code A9

Introduction:

Glass to metal seal damage is very important as gas, vapour, dust or even water could impregnate the fitting through the damage. This seal should never be bodged up. If possible always replace with a manufacturer's seal or buy a new unit.

Has the Instrument got a damaged Metal – Glass Seal?

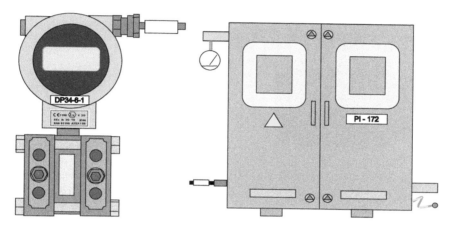

If the instrument has a see through display as in the diagram top left, ensure that the seal is still intact where the clear display meets the metal case. There have been instances where the seal has been damaged and water has got into the enclosure. An instrument similar to the one in the diagram may cost several hundred pounds.

Sometimes it is an instrument cabinet with clear glass windows similar to the diagram (top right). These cabinets are sometimes pressurised Exp (see next Section) and sometimes just an instrument cabinet. Unless the cabinet is Exp, and it is an instrument cabinet, it is unlikely to contain anything that would produce heat or hot sparks. Report any defects you find on your Inspection.

In some cases Inspectors will not discover any damage until they come to the **Detailed** Inspection.

# Visual Inspection
## Are All of the Glands the Correct Protection & Complete?
## (Gland Protection)

## Code A10

### Introduction:

On a **Visual** Inspection, what are Inspectors looking for in the way of glands? Well firstly are there any uncertified glands in Certified equipment? Does anything look wrong with the gland? Is the gland 'Protection' correct for the protection of the equipment into which it is fitted, e.g. Exd glands in Exd equipment etc. All the Inspector has at the moment is an **Area Classification Drawing.**

### Dates:

 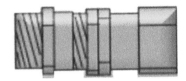

Let us go back to **before 2003**, at this time a **ONE SEAL UNCERTIFIED** gland could be used in Exe increased safety equipment, Exn reduced risk and Exi intrinsic safety quite legally and there will still be thousands out there in the field.

**From 2003 – 2008** it was legal in all of the above protections to use a **TWO SEAL UNCERTIFIED** gland, but a **ONE SEAL UNCERTIFIED** gland must no longer be used.

From **2008 to the present day,** the rule is that a correct **CERTIFIED** gland be used in **CERTIFIED** equipment. Universal glands are ideal. If the uncertified gland (Pre-2008) has to be removed from the equipment **FOR ANY REASON** e.g. rewiring or maintenance **THEN A CERTIFIED GLAND MUST GO BACK.**

So to sum up gland protection, it must be a correct Certified gland in Certified equipment and on this **Visual** Inspection no tools may be used to move the gland, the protection markings must be read with the gland in place without turning it. When adaptors or reducers are used they also must be of the correct protection.

# Visual Inspection
## Are All of the Glands the Correct Protection & Complete? (Gland Protection)

## Code A10-1

### Compression Glands:

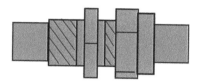

Compression glands (technicians call them 'stuffing' glands) are used when the cable does not have a steel wire armour or braid. A normal universal gland would work, but it would be a huge waste of money on any large project. These metal glands can be Atex Certified: Exd, Exe, ExnR & Ext. **The Plastic Compression Glands will never be Exd.**

### Gland fitted incorrectly (SWA Showing):

Can you see steel wire armour (SWA) protruding out the back of the gland? If so the sheath seal will be ineffective. Water, gas or vapour can enter here and may bypass the deluge seal if one is fitted, leaving only the inner seal of the two seal gland to be effective. This of course is usually bad workmanship and may not be quite as blatant as the diagram on the left, so look carefully.

### Gland fitted incorrectly (Seal Showing):

The sheath seal may be faulty and protruding out of the top of the gland as in the diagram on the right. This situation, like the example above, will mean an ineffective sheath seal and gas or water can then bypass this section of the gland, so again you are relying purely on the inner seal. To sum up, there would, in effect, only be a **ONE SEAL** gland in this case as the sheath seal is ineffective.

### Reducers fitted incorrectly:

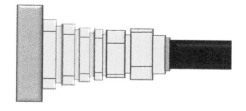

When reducers are used you can only use **ONE REDUCER PER GLAND**, not a Christmas tree of them as shown left. There is a case that the volume could be increased with a large number of reducers.

# Visual Inspection
## Are All Stoppers Present and the Correct Protection?

## Code A11-6

### Introduction:

On a **Visual** Inspection the Inspector would check the equipment to see if any stoppers were missing. The next check should be to see if the stoppers that are in place are the correct 'Protection' for the equipment e.g. Exd flameproof stoppers in Exd flameproof equipment. All the Inspector would have at this point is an **Area Classification Drawing.**

### Stopper Correct IP:

The stopper to the right is a brass/metal stopper with an 'O' ring so that when it is installed this ring helps to seal into the equipment. When the stopper is in place you may not see this 'O' ring.

### Stopper Metals:

The stopper does not have to be brass. The stopper to the left is stainless steel.

Again you will notice the 'O' ring.

### Stopper correct Protection:

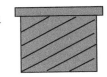

The stopper **must** be of the correct 'Protection' for the equipment e.g. an Exd stopper in Exd equipment and an Exe stopper in Exe equipment.

Stoppers are usually screwed into the equipment from the outside, but it is possible to have stoppers screwed in from the inside. These are usually on equipment like bus chambers of lighting panels. This would force you to isolate before removing the stopper.

### Black Plastic Stoppers:

In many cases, although not always, Exe stoppers are made of black plastic as shown left. Under **no** circumstances must these be used in Exd flameproof equipment. These stoppers have an 'O' ring to ensure the IP in Exe, Exn, Exi & Ext.

# Visual Inspection
## Are All Stoppers Present and the Correct Protection?

### Code A10-3

### Stoppers fitted into Reducers:

Stoppers must **never** be fitted and used in reducers. If there is a 25mm hole I suppose that it would look easy to get a 25mm – 20mm reducer and use a 20mm stopper. The correct size stopper must be used for the size of the hole.

### Transit Stoppers:

Transit stoppers are just what they say, they stop dirt and dust getting into the equipment in transit and while it sits on a stores shelf. These are usually very bright colours such as yellow or red. This temporary stopper can be screwed, a push fit or simply a sticky label over the hole. They must be removed and the correct Certified stopper used.

### Completely Parallel Stoppers:

It is unlikely that there will be any of this type stopper (left) still in use, but very old stoppers have been known to be parallel with no rim, so if you kept on tightening it would drop into the equipment. These would have to be fitted with a locknut to lock the stopper in place.

### IS Junction Boxes:

If a junction box is used in an IS loop and there is only one loop then the junction box need **not** be Certified and classed as **'Simple Apparatus'** and can be a biscuit tin (so long as the IP is suitable). If the box has more than one loop then the box must be Certified e.g. Exe. The box then becomes not quite so simple as for one loop. If the box is Certified Exe then Exe glands & stoppers must be used.

### Certification:

On a Visual Inspection you should be able to read the Certification with the stopper in place. If you cannot see any markings then you assume that it is not Certified.

# Close Inspection
## Are all of the Gland Sections tight?

### Introduction:

On a **Close** Inspection the Inspector can now get hold of the glands and see if they are loose. Inspectors can also determine if the gland should be a standard gland or a barrier gland because they have more documentation here.

### Gland Sections Loose:

Ensure all gland sections are tight. Inspectors should not be able to undo the gland by hand, not even the knurled section if they are that type.

### Gland fitted incorrectly (SWA Showing):

Can you see steel wire armour (SWA) protruding out the back of the gland? If so, the sheath seal will be ineffective. Water, gas or vapour can enter here and may bypass the deluge seal if one is fitted, leaving only the inner seal of the two seal gland to be effective. Even if the gland was tightened this would still be a problem. This of course is usually bad workmanship and may not be quite as blatant as the diagram on the left so look carefully.

### Gland fitted incorrectly (Seal Showing):

The sheath seal may be faulty and protruding out of the top of the gland as in the diagram on the right. This situation, like the example above, will mean an ineffective sheath seal and gas, water or dust can then bypass this section of the gland so again you are relying purely on the inner seal. To sum this up, you only have a **ONE SEAL** gland in this case as the sheath seal is ineffective. Even if the gland was tight this would be a problem.

# Close Inspection
## Are All Stoppers Tight?

## Code A10-5

### Introduction:

On a **Visual** Inspection the Inspector would be looking to see that the stopper is **present,** has the correct protection and is correct for the equipment. With a **Close** Inspection Inspectors are checking whether the stopper is **tight.**

### Loose Stoppers Exd:

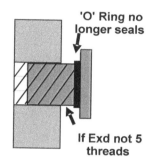

Loose stoppers are a hazard and, depending upon the vibration level, could vibrate out of the equipment completely.

If the stopper is in Exd flameproof equipment, coming loose may mean it no longer has the correct number of threads entered.

Very old stoppers have been known to be parallel with no rim, so if you kept on tightening it would drop into the equipment. These would have to be fitted with a locknut to lock the stopper in place.

### Loose Stoppers Exe, Exn & Exi:

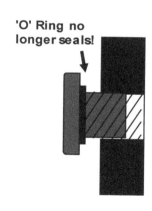

If the stopper is in Exe increased safety, Exn reduced risk, Exi intrinsically safe, or Ext protection by enclosure equipment, it coming loose may mean not maintaining the correct IP seal.

For this equipment the minimum IP is IP54 and unless the 'O' ring on the stopper is tight up against the equipment then the ingress protection of that equipment could be compromised and water, gas, vapour or dust could get in.

### Loose Stoppers Restricted Breathing:

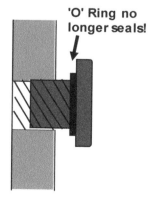

If the stopper is in ExnR restricted breathing equipment, it coming loose may mean not maintaining the correct IP seal.

For this equipment the minimum IP is IP54 and unless the 'O' ring on the stopper is tight up against the equipment then the ingress protection of that equipment could be compromised and water, gas or vapour could get in.

The restricted breathing of the fitting would be non-existent.

# Detailed Inspection
## Is there any Damage or Unauthorised Modifications Inside the Equipment?

## Code A11

### Introduction:

On a **Visual** Inspection the outside of the equipment is all that can be checked for damage, but now that the equipment has been opened up on the **Detailed** Inspection, flange faces, screw threads & gaskets can be checked.

### Are all Instrument Screw Threads Clean & Undamaged:

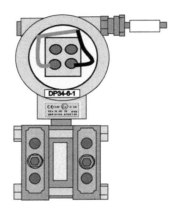

Ensure that all of the instrument screw lid threads are clean and free from damage. Even the slightest damage should be reported as the instrument IP could be compromised. Do not forget to check the lid as well as the main body. On closing the enclosure ensure that a smear of non-setting grease is applied to the threads (if applicable) both on the equipment and on the lid. The lid should be screwed back to its correct position.

### Is the Terminal Block Undamaged and the correct Manufacturer's?

Ensure that the terminal block is the manufacturer's and the correct one for the equipment, not an unauthorised strip terminal block. The terminal block should be fastened into the equipment as per Manufacturer's Instructions which, with an instrument similar to the one above, it will probably be part of the instrument itself and not a separate item. Check the wiring for damage around the block.

### If Applicable, Is the Gasket/O Ring in good condition?

Sometimes instrument equipment has a gasket or an O ring which could get worn or fragmented which would of course make the IP questionable. This gasket/O ring should be a manufacturer's gasket. Another most important point is that a gasket or O ring must **never** be left off equipment if it is meant to have one as this could affect the IP and instrument integrity.

# Visual Inspection

## Is there any evidence of Unauthorised Modifications Outside the Equipment?

## Code A12

Introduction:

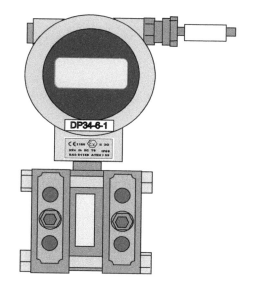

One of the problems is that if we take an example of an instrument similar to the one shown left, the manufacturer states that the equipment is installed in a particular way. They stipulate what glands, stoppers and IP protection to use. If there is an explosion, or the instrument suffered severe damage in some way, and it was found that this equipment was the cause and the technician had installed it **EXACTLY** to the manufacturer's specification, the explosion would be the **MANUFACTURER'S** fault. If the equipment had not been installed **EXACTLY** as per specification and the technician had carried out an unauthorised modification, which might be using different IP washers, or the wrong stoppers etc, in court the manufacturer could put some or all of the blame on the way it was installed so it would be the **TECHNICIAN'S** fault.

So **NEVER** modify certified equipment. If you do the Certificate Conformity becomes null & void. People do things with the best of intentions, but on the face of it some are quite dangerous. Let us look at several examples below, I am sure you can think of more.

Unauthorised Modifications might include:

1   Bolts, Grub Screws and screws that are not the manufacturer's.

2   Anything that looks as if it were added to the manufacturer's specification and you suspect this might not have the manufacturer's approval.

3   Anything that is obviously missed off and you think it is without the manufacturer's approval.

4   Incorrect/uncertified glands or stoppers.

# Visual Inspection

## Are Safety Barriers, Galvanic Isolators, Relays and other Energy Limiting Devices of the Approved Type, Installed in Accordance with the Certification Requirements and Securely Earthed Where Required?

## Code A13

### Introduction:

If barriers are not installed exactly as per their specification then they may not work and not protect the loop at all. On a **Visual** Inspection, Inspectors do not have a Loop Diagram so that they cannot check if any details or numbers are correct on the barrier, but some **Visual** faults are obvious as follows:

### Are all Barriers the correct way up?

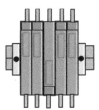

If there is a Zener barrier upside down, the thing is, it will actually still work, but this action will bypass the barrier fuse and would be relying on some distribution fuse possibly feeding the 28V system. The up-line fuse may not be fast enough for the barrier protection and, for a fraction of a second, allow above IS voltage onto site.

### Are Zener & Galvanic Barriers on Separate Rails?

Zener and Galvanic barriers should be on different rails because of the different earthing arrangement of the barriers. Galvanic barrier screens for instance may not connect to the clean earth!

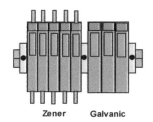

Zener          Galvanic

### Are Zener Barriers Earthed Correctly?

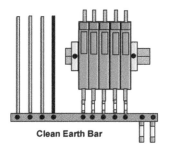

Clean Earth Bar

Zener barriers, as above, require a very high integrity earth of **1Ω**. If Inspectors were to take a Zener Barrier Loop Diagram and look at what would happen if the earth was removed, they would see that the barrier just would not work and would fail to protect the loop in the case of a fault. So not earthing the Zener barriers properly is extremely dangerous.

# Detailed Inspection
## Condition of Enclosure Gasket Satisfactory?

## Code A14

### Introduction:

On a **Visual** Inspection the outside of the equipment is all that can be checked for damage, but now that the equipment has been opened up on the **Detailed** Inspection screw threads, seals & gaskets can be checked. If the seal/gasket is faulty then water, dust, gases & vapours may be able to enter the enclosure and set up a very dangerous situation.

### Is the Exe Increased Safety Enclosure Gasket satisfactory?

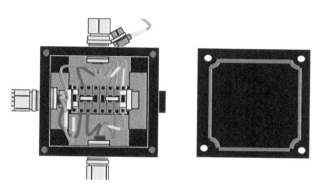

Sometimes when Inspections are carried out Inspectors remove the enclosure lid and put it on the floor whilst they inspect the enclosure.

On completion of their inspection they pick up the enclosure lid and replace it without inspecting the lid itself to see if the gasket is intact or if anything is amiss.

### Is the Exi/Exe Enclosure Gasket satisfactory?

In the type of enclosure shown on the right the seal/gasket is usually fastened to the lid itself. Sometimes it is around the rim of the enclosure. Any parts of the seal/gasket that are damaged or missing will definitely compromise the IP of the box. Although with IS there is no problem with sparking, water could cause the IS Loop to shutdown.

If the seal is compromised then gas, vapour or dust could also enter the enclosure. Exe enclosures could be a higher voltage.

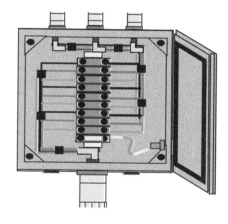

# Detailed Inspection
## Are All of the Electrical Connections Tight Inside the Enclosure?

### Introduction:

Now that the cover has been removed the screws inside of the equipment must be checked for tightness. Remember it might be years before another **Detailed** Inspection is carried out.

### Terminal Screws:

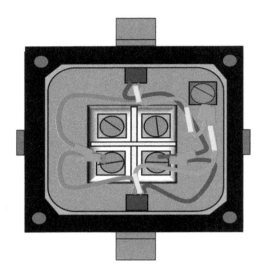

Now the equipment has been opened up the terminal screws inside must be checked for tightness. Any spare screws in unused terminals are also tightened at this stage to stop them falling out due to vibration and maybe shorting out terminals below. Depending upon the type of equipment, if any screws are loose this could cause heat. If the equipment is Exe increased safety, this would not be a good thing. If the equipment is Exi, heat would be unlikely but malfunction possible.

### Marshalling Boxes:

Admittedly being Intrinsically Safe there would be no problem with heat or sparks, but loose connections may well affect the IS Loop and, of course, the running of the plant.

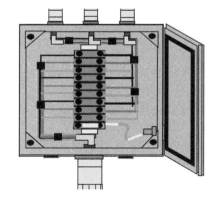

Again, although an Intrinsically Safe System it is best to complete this exercise with the Loop dead in case it affects the running of the plant.

Company policy would prevail here.

# Detailed Inspection
## Are Printed Circuit Boards Clean and Undamaged?

## Code A16

### Introduction:

Many instruments have printed circuit boards these days and these are part of the Instrument Certification. Any damage to the circuit board would put the instrument function in jeopardy. Inspect the boards carefully.

### The Printed Circuit Board:

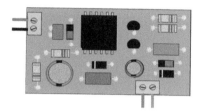

A printed circuit board is a way of connecting a whole range of components in a circuit by a copper connecting system fastened to a board and components soldered on. It could be described as a miniature 'Bus' system, saving messy wiring connections.

### Water Damage:

Water damage can be very devastating for the printed circuit board and is usually seen as a yellowish tinge on the green board. Even storing the board in humid atmosphere conditions can harm the unit.

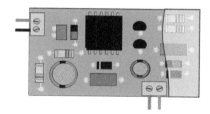

### Components just burning out:

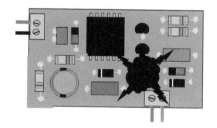

If components just burn out e.g. a capacitor (some go with quite a bang), these are the easiest to spot with black emanating outwards from the faulty component. Repairs must be made by a reputable Certified company and must be brought back to manufacturer's Standard. (Or buy a new board.)

### Circuit Boards Destroyed by Static & Smothering:

Conditions which are high in static charge can have a devastating effect on a printed circuit board. Let us take areas where there is highly charged static dust. Dust can also stop heat dispersion so individual components can become overheated and burn out.

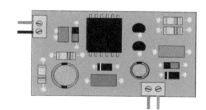

# Close Inspection
## Ensure that the Maximum Voltage Um is Not Exceeded.

## Code A17

### Introduction:

The Um is the maximum RMS voltage that can be applied to say, the Non-Intrinsically Safe connections of an interface unit such as a barrier without damage or invalidating documentation.

### Intrinsically Safe Parameters:

Let us take the interface unit as a barrier; we have mentioned Um above. The Barrier input of a 1.5 safety factor barrier [ llC ] to llC equipment on site is between 28volts & 42volts. Because of the 300Ω current limiting resistor in the barrier, the output current is around 93mA. As well as Um you might also be interested in the definitions below:

### Non Hazardous Area:

| Interface: | Safety Description Parameters: |
|------------|-------------------------------|
| Uo | Maximum Output Voltage under fault conditions |
| Io | Maximum Output Current under fault conditions |
| Po | Maximum Output Power under fault conditions |
| Co | Circuit Capacitance |
| Lo | Circuit inductance |

### Site:

| Site: | Safety Description Parameters: |
|-------|-------------------------------|
| Ui | Instrument Input Volts |
| Ii | Instrument Input Current |
| Pi | Instrument Input Power |
| Ci | Instrument Capacitance |
| Li | Instrument Inductance |

# Visual Inspection
## Cables are Installed in Accordance with the Documentation?

## Code B1

### Introduction:

When we talk about Intrinsic Safety, care must be taken to ensure that the cables have been installed as per documentation and the installation is safe and to IS specifications. The diagram below shows how the cables might be arranged in a barrier enclosure, i.e. spare cores, screens, dirty & clean earths etc. Remember you are limited with documentation on a **Visual** Inspection according to Standards to an **Area Classification Drawing not a Loop Diagram.**

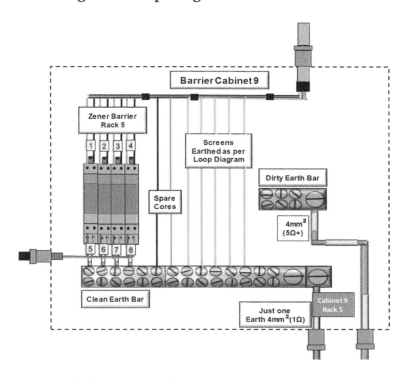

So some of the things you would be looking for in cabling:

1   Are all the cables blue, if not, are they marked?

2   Are all cables connected somewhere?

3   Is the dirty earth separate from the clean earth?

4   Are earth bars marked 'clean' and 'dirty'?

5   Are input and output cable sheaths marked?

6   Are the barrier input and output cables pairs/cores marked?

# Detailed Inspection
## Cable Screens are Earthed in Accordance with Documentation

Code B2

Introduction:

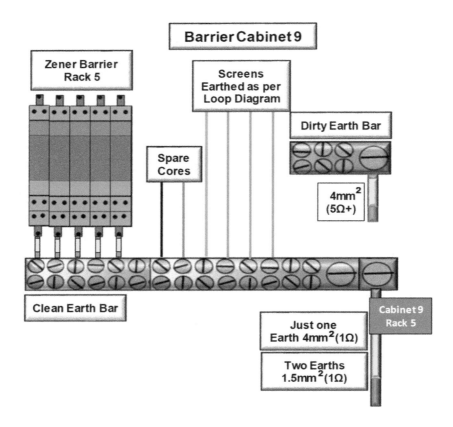

The screen of a cable is not, as some people think, the bare wires that come out of the end where we put on green or white sleeving. The screen is the insulated foil that runs throughout the cable and surrounds the conductors. Sometimes there is a foil screen around each pair of a multi-core, and a further foil screen around all of the pairs. The bare wire that comes out of the end is in contact with the conducting side of the foil all the way through the cable, so by earthing this you earth the foil screen.

This achieves the prevention of the invasion of other voltages by surrounding the pairs in a Faraday Cage. Therefore, Inspectors must look at the Loop Diagram and see where the screens are connected i.e. dirty or clean earth, and ensure that the equipment matches the Diagram. **Screens are connected as per Loop Diagram.**

# Visual Inspection
## Any Damage to Cables?

## Code B3

### Introduction:

When I mention here 'damage to cables' on a **Visual** Inspection, the Inspectors would need to know in their remit as to just how far up the line do they go? Inspectors cannot be expected to examine every metre of cable in great detail on, say, a 500 metre run, so a certain length from the equipment will be inspected in detail. The Inspector may walk the entire run to see if there is any conspicuous catastrophic damage.

### Is there any local damage to IS Cables?

The cable trauma is thankfully not usually as devastating as the diagram above. Intrinsically Safe cables (Light Blue in majority of cases) showing damage should again be reported immediately. Although probably not dangerous as far as sparking and ignition is concerned, this type of damage could cause instruments to malfunction, the plant to go into shutdown etc. Can Inspectors see any damage to the cables, cable tray or whatever cable management is being used e.g. ladder rack, conduit, trunking etc. They may find damage with steel wire armour showing or even conductors in extreme cases as above. On finding any catastrophic cable damage this must be reported to the control room and engineer IMMEDIATELY, not just noted down in some documentation.

### Is there any damage to cable at the Gland?

The situation illustrated on the diagram to the right might be more common where the steel wire armour or braid is showing at the gland. This might be because someone has not made the gland off properly in the first place.

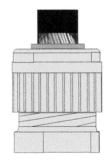

If the cable is showing steel wire armour (SWA) or braid on a vertical run it must be reported straight away as water can enter here and bypass the outer sheath seal on the gland.

# Visual Inspection
## Is the Sealing of Trunking, Ducts, Pipes & Conduits Satisfactory?

## Code B4

### Introduction:

Ensure that ducts & pipes are sealed. The sealing of the trunking/duct does not in fact come under Atex as they are classed as what is called **'Non-Sparking'**

### Sealing trunking/ducts in Hazardous Areas:

The cables come through the duct as shown left and a manufacturer's resin is injected in around the cable to make a seal as per instructions.

So the company to whom the ducting belongs to would end up with a **'Statement of Exclusion'.**

What the seal achieves is the prevention of the migration of water and/or gas into the switch room or substation. The sealant can be fire proof and also act as a fire barrier, as well as being vermin proof.

### Stopper boxes for conduits:

Ensure that stopper boxes are fitted in conduits where appropriate. Stopper boxes are compound filled 'tubes' that fit into the conduit system shown as in the picture, right. These are filled with manufacturer's compound and are designed to stop explosions passing along the conduit.

In my experience, conduit systems are not common in hazardous areas these days, but if they are used then there should be a stopper box in the conduit where it leaves the hazardous area to stop explosions passing into a non-hazardous area.

# Detailed Inspection
## Are Point to Point Connections Correct? (Initial Inspection only.)

Code B5

Introduction:

The Point to Point Check should have been completed on the **Initial Inspection.** This should not have to be done now unless major changes have been completed on the loop wiring.

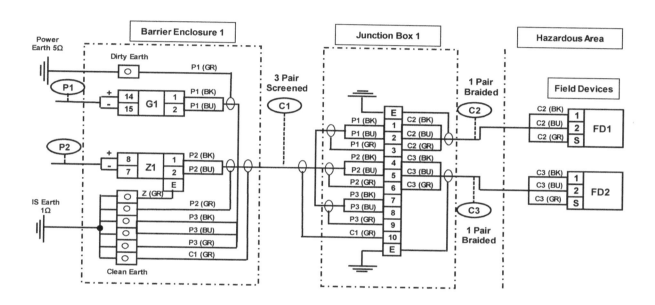

What Inspectors are actually checking is that the cables in the loop are exactly how they are portrayed on the Loop Diagram all the way from the barrier units to the instruments on site. Remember, these cables could be tens of metres long and could go through several junction boxes/marshalling boxes. Also remember that, as mentioned above, this is **only** done on the **initial inspection.** How this is done depends upon company commissioning policy. If there are any major alterations to the wiring then Point to Point checks may have to be repeated.

# Detailed Inspection
## Is the Earth Continuity Satisfactory?
## (Connections Tight, Cross Section of Conductor)

Code B6

Introduction:

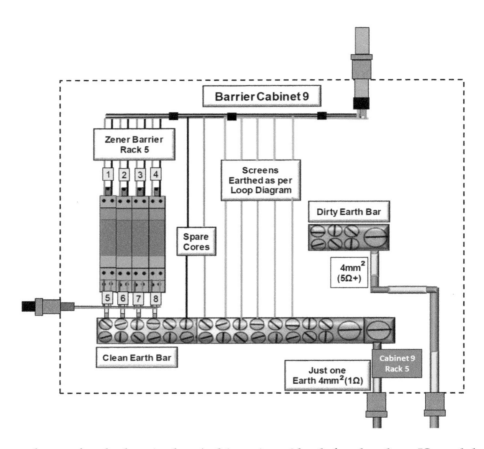

Is the earthing conductor for the barrier box/cabinet 4mm$^2$ both for the clean IS earth bar and the dirty earth bar? Are all of the non-galvanic i.e. Zener barriers, earthed to this clean earth bar? Are the connections tight on both earth bars and earth continuity satisfactory? Ensure that if you use a continuity tester that is **approved for working on IS circuits,** that it also carries out IR tests at 500 volts.

**DO NOT PRESS MEG-OHMS WITH THE INSTRUMENTS AND BARRIER UNITS CONNECTED.**

It is arguable that the cross section of the earth wire on a small system could be picked up even on a Visual or Close Inspection without waiting until the Detailed.

# Detailed Inspection
## Do Earth Connections Maintain the Integrity of the Type of Protection?

## Code B7

### Introduction:

The IS earth is tested and the reading compared to the **Initial** Inspection from when the system was first installed. A dynamic graph of the readings can be formed and on each period when the earth reading was taken, the readings can be entered and the graph will automatically update and show any degradation.

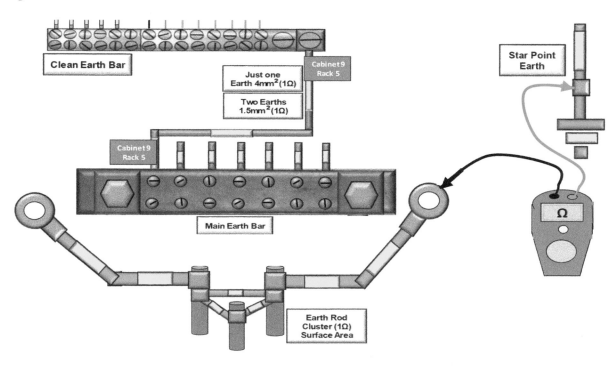

If it is the main **1Ω** earth rod cluster that is under test this would have to be done under **PLANT SHUTDOWN CONDITIONS WITH ALL LOOPS NOT IN USE.**

Testing between the cluster and a known 1Ω earth would be required i.e. the star point of the distribution transformer, or a full earth rod test as described in Power Section B7 Earth Resistance of a TT System **(Engineers Decision).**

**Do not disconnect the earth rod on the transformer star point under any circumstances.**

**Disconnecting the IS earth bar as above will leave all clean earths from that bar 'Floating'.**

# Detailed Inspection
## Is the Intrinsically Safe Earthing Satisfactory?

## Code B8

Introduction:

The system should be securely earthed to a **1Ω** earth rod. Check the connection to the rod by removing the earth from the main earth rod/bar and the system should be earth free. **This test should only be carried out with the system SHUTDOWN.**

IS Earthing System:

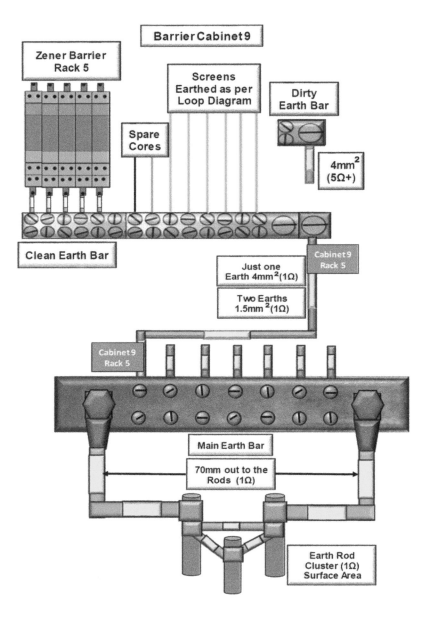

# Detailed Inspection
## Is the Insulation Resistance (IR) Satisfactory?

### Code B9

### Introduction:

With this IR test on instrumentation it is very important that, for example, a differential pressure transmitter, is **DISCONNECTED**, and that any barriers are also **DISCONNECTED.** The insulation tester or Megger as technicians call it, will be set to meg-ohms **(MΩ)** and the voltage here is **500 volts**. All you are testing really is the cable.

### IR Testing a Multicore:

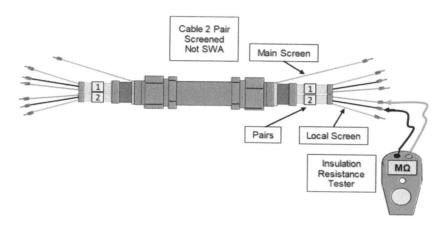

### Example of Testing:

1   Disconnect any Instruments and Barrier Units.

2   Put Warning Notices on the other end of the Cable to the Testing.

3   Testing from outside in, you will need a Gas Free Certificate for the Test Instrument.

4   Testing from inside out, the other end will need to be connected in a suitable enclosure.

5   Set the test Instrument to MΩ 500 Volts.

6   Test between the cores of each pair.

7   Test between each core and local screen.

8   Test between local screens and main screen.

9   Test between cores and main screen.

10   The reading should be quite high - the more towards ∞ the better.

# Detailed Inspection
## Is there at least 50mm between IS and Non-IS in Distribution Boxes & Relay Cubicles?

### Code B10

### Introduction:

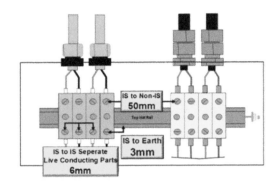

As you can see in the above diagram the clearance distance between IS and non-IS terminals is 50mm which in old measurements is around 1.5 inches. The clearance distance between IS live conducting parts is 6mm and the clearance distance between IS and earth is 3mm. These last two distances, as shown in the diagram above and in most cases, will be calculated for you by the terminal block.

### Clearance Distance:

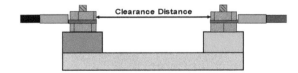

The Standard does in fact state **'Clearance'** distance as above and not **'Creepage'** distance as below so what is the difference? The Clearance distance is the distance through air between two points.

### Creepage Distance:

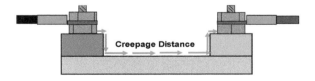

The Creepage distance is the shortest distance a track would take along a surface of an insulator between two conducting terminals.

# Detailed Inspection

## Is the Short Circuit Protection of the Power Supply in Accordance with Documentation?

Code B11

Introduction:

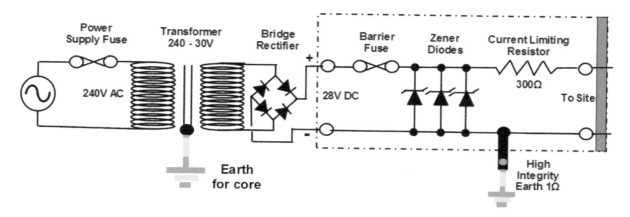

Illustrated above is a very simple circuit of a Zener barrier and a possible power supply. Look at the non-intrinsically safe end of the barrier and decide whether it would be possible, in certain circumstances, for these terminals be fed by a voltage that is greater than the Um of the Barrier.

You see that we use a double wound transformer with an earth core. If the primary winding melted it would have to go through the earth core and blow the fuse before it could melt into the secondary, which, if we did not take precautions, could cause the voltage to rise significantly.

This is why transformers such as auto-transformers must **never** be used, as a fault here could cause the full input voltage to be transferred to the output.

Transformer galvanic interface units use isolating transformers for their input voltages for the safe and hazardous area side. This, of course, would be called a 'floating' supply with no reference to earth.

# Detailed Inspection
## Are All Conditions of Use Complied With?
## (Items with an 'X' after the Atex/BASEEFA Number)

### Code B12

Introduction:

This section refers to a scenario where the equipment has an **'X'** after the Atex number. The **'X'** refers to **'Special Conditions of Use',** which will also be stated on the Certificate which comes with the equipment. The Inspector must look at the Certificate, note the Special Conditions, and see that they are being complied with.

Where is the 'X'?

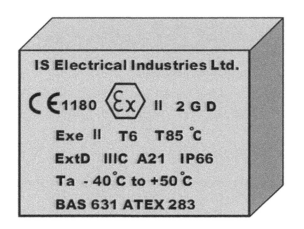

As you can see on the diagram to the left the 'X' follows the Atex or BASEEFA number. What Inspectors are looking for now is the Certification that came with the equipment so that they can see what the Special Conditions of Use are, and then see that they are being complied with.

It is very important that the Installer is shown the Certification so that they are aware of what Special Conditions they have to abide by when installing the equipment.

Three examples of Special Conditions:

| Enclosures assigned an impact level of 4J or 4NM and must be installed only in areas of low mechanical danger | Precautions must be taken to ensure that the thickness of dust layer on the terminal box will not exceed 5mm | Wiring within the enclosure must not be grouped or bunched to prevent hot spots forming |

# Detailed Inspection
## Are Cables Not in Use Correctly Terminated?

Code B13

Introduction:

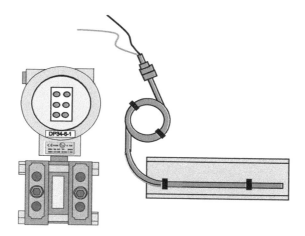

When an item of equipment is disconnected in a hazardous area, perhaps in preparation for removal, then the disconnected cables must not be left hanging on site. They must be terminated in an appropriate enclosure and labelled with isolation details and, if applicable, a Permit number.

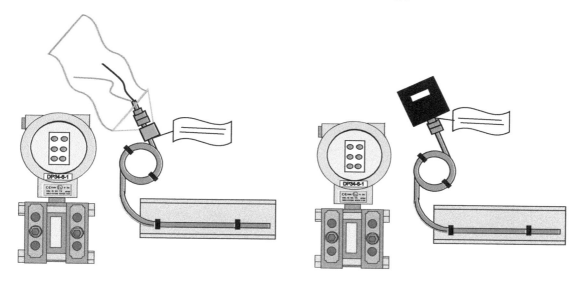

The above two diagrams illustrate ways the cable can be protected after disconnection. Above left, the cable end is put into a plastic bag with tape around the bottom to prevent damp and a label attached giving instrument tag number, Permit details etc. Above right, the cable is terminated into a correct Certified box for the area. It depends upon company procedure & policy.

# Visual Inspection
## Is the Equipment Protected Against the Weather?

## Code C1

### Introduction:

We cannot put our hazardous area electrical and instrument equipment inside a bubble and protect it from all weathers. The equipment ingress protection (IP) will usually protect it against solids and liquid ingress. What about direct sunlight and, at the other end of the scale, extreme cold? Direct sunlight can obviously cause the equipment to heat up.

Older equipment has a temperature classification that usually requires the normal ambient temperature to be between **-20C to +40C**. More modern equipment has higher and lower ambient temperatures, but might be worth investigation. Another point to consider is that at very low and very high temperatures, certain electrical insulation materials will degrade and, to some extent, lose their insulating property. If the ambient temperatures are known to be higher or lower than the 'norm', equipment with Ta or Tamb can be obtained with a higher or lower ambient temperature.

### Very Low Temperatures:

The inspector(s) should be concerned as to whether in very low temperatures the hazardous area equipment can maintain the 'Protection' that make it a Certified item i.e. Exd, Exe, Exi, by way of the material used to construct the equipment, the norm being -20C to +40C. So:

1   **Can the equipment function efficiently at very low temperatures, for instance if the temperature was to go below the norm of -20C?**

2   **Could the surrounding atmosphere gas/vapour change its properties at very low temperatures?**

3   **Can the equipment function efficiently at very high temperatures, for instance if the temperature was to go over the norm of +40C?**

### High Temperatures (e.g. Direct Sunlight):

Heat can cause electrical insulation degeneration. Cables in conduits which are in direct sunlight can be subjected to very high temperatures. PVC conduits & pipes can become more brittle when subjected to direct sunlight due to ultra violet (UV) light on the PVC. Heat can build up outside and inside enclosures, especially if they are made of metal. Manufacturers may give guidance on ambient temperatures.

# Visual Inspection

## Is the Equipment Protected Against Corrosion?

## Code C1-1

### Introduction:

Corrosion is usually caused by some kind of chemical reaction. In many cases it is because the equipment is the **'Anode'**. By using **'Cathodic Protection'** we can cause the equipment to be a **'Cathode'** instead. This can slow the corrosion down enormously. Below, we talk about looking for corrosion on Inspections and what signs the Inspectors might be looking for.

### Inspecting Equipment for Corrosion:

Inspectors carry out the Inspection, if there is any corrosion they must Risk Assess:

1   **What is the corrosion and what equipment is it on?**

2   **Is the corrosion minimal and requires no action?**

3   **Is the corrosion acceptable at the time of Inspection but a Risk Assessment determines it will get worse in the near future?**

4   **Is the corrosion great at the time of the Inspection and requires maintenance work straight away?**

### What is Corrosion?

I always look at corrosion as being released energy that was trapped into, say, steel at its liquid state and the metal is now returning to a more stable state e.g. Oxides, Sulphides or Hydroxides. If we look at any rusty structure that is made of steel, for example a rusty pipeline without any cathodic protection, it is not all rust (Ferrous Oxide), there are some good areas here and there. Where those good and bad areas meet, a cell is formed between the good parts (Cathodes) and the corroding parts (Anodes), the electrolyte (Water) provided by moist air or soil.

### Cathodic Protection:

We can use cathodic protection on a pipeline to slow this process down, explained in the book **'Hazardous Areas for Technicians' ISBN: 978-1-912014-95-8**

| Cathodic | |
|---|---|
| ↑ | Platinum |
| | Gold |
| | Titanium |
| | Silver |
| | Tin |
| | Lead |
| | Mild steel |
| | Aluminium |
| Anodic ↓ | Zinc |

If we take a cable tray, the galvanising is a type of zinc coating cathodic protection. If you look at the cathodic table left you will notice that zinc is more **'Anodic'** than mild steel hence the zinc forces the mild steel to be a **'Cathode'**. Corrosion will always go for the anode i.e. the Zinc. If you look at Gold you will see that it is near the top of the table i.e. very cathodic, which is why you can dig up gold coins out of the ground and there is hardly any corrosion at all.

# Visual inspection
## Is the Equipment Protected Against Vibration?

Introduction:

There are many things that can cause vibration in chemical factories and on platforms. In the electrical world, electric machines spring straight to mind e.g. motors, generators etc. In the mechanical world there are compressors, pumps etc. In the process world there are the plant processes themselves. Vibration causes loose threads, bad connections and cracking where there is a metal seam.

Vibration effects on several instruments:

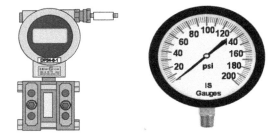

There are several instruments on which vibration may have a very adverse effect, let us look at a few:

1   Pressure Sensors: vibration can severely disturb the measurement signal as it can be transmitted to the output signal.

2   Transmitters: very susceptible to prolonged vibration. Most manufacturers will recommend that they be mounted, if possible, in areas with the lowest vibration.

3   Load Cells: are susceptible to vibration and again this can be transferred to the output and give inaccurate readings.

4   Pressure Gauges: Pressure gauges, by the mere fact of their mechanisms, are very susceptible to vibration. Oil filled gauges are less likely to fail due to vibration than a standard pressure gauge.

Where possible, instruments should be mounted in areas of least vibration, however, saying that is easy, carrying it out on plant may be extremely difficult. The location of the instrument depends upon the process. It may be an idea to use some vibration monitoring equipment in the event of several failures, and obtain some evidence of exactly how much vibration there is.

# Visual Inspection
## Is There Accumulation of Dust or Dirt?

## Code C2

## Introduction:

If dust or dirt is allowed to build up on the equipment then this could have a detrimental effect on the temperature of the equipment. On a **Visual** Inspection you are just looking to see if there is a build-up of dust or dirt on the equipment.

## Temperature Classification in General:

| | |
|---|---|
| T1 | 450 |
| T2 | 300 |
| T3 | 200 |
| T4 | 135 |
| T5 | 100 |
| T6 | 85 |

The table to the left shows the six temperature classes in the UK. Equipment should not go over these temperatures under **NORMAL AND SPECIFIED FAULT CONDITIONS.** The ambient temperature for these temperature classes has a norm of **-20C to +40C**

If the Ambient Temperature is higher or lower then alternative equipment must be obtained, in which case the data plate will state **Ta** or **Tamb** and the new higher and/or lower ambient temperature will be stated.

## How Hot are these Temperatures?

These temperatures are taken for granted, but do you realise how hot the equipment would be if it reached 450C as in T1? Look at the effects that temperature has on certain materials, some of which are used in the electrical world,

In the table on the right you may be able to get some idea of what can really happen if things start to heat up due to fault conditions.

| | |
|---|---|
| 455 | PVC Ignites |
| 327 | Lead Melts |
| 232 | Tin Melts |
| 135 | Polyethylene Melts |
| 100 | Water Boils |
| 85 | Burn Hands |

WE DO NOT WANT ANY CONDITIONS THAT MAY CAUSE THE EQUIPMENT TEMPERATURE TO RISE, AND EQUIPMENT COVERED IN DIRT AND DEBRIS IS ONE OF THEM.

# Visual Inspection
## Is There Accumulation of Dust or Dirt?

## Code C2-1

### Dust Ignition Temperatures:

If we just work on a golden rule that 'clouds explode and layers burn', the table (right) shows the ignition temperatures of various dusts with the different ignition temperatures of their clouds and layers. Our equipment temperature must not get anywhere near to these temperatures or there could be an explosion (Cloud) or fire (Layer). Excessive dust or dirt beyond 5mm could smother the equipment and cause it to heat up so good housekeeping is vital.

| Dust | Ignition Temperatures °C | |
| --- | --- | --- |
| | Cloud | Layer |
| Cellulose | 520 | 410 |
| Flour | 510 | 300 |
| Grain | 510 | 300 |
| Sugar Dust | 490 | 460 |
| Tea Dust | 490 | 340 |
| Starch | 460 | 435 |
| Lignite | 390 | 225 |

### Dust & Dirt on IS equipment:

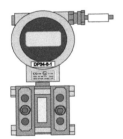

Dust & Dirt on IS Instruments will not usually cause heat as there is no real power there. What it may cause, especially in the case of a transmitter, is for the instrument to malfunction. For example sometimes an instrument takes its reference from atmospheric pressure. If any debris clogged up the reference part of the instrument it would malfunction. Dust & dirt on metal equipment can also cause corrosion so good housekeeping is essential.

### Dust & Dirt on Exp Pressurised Equipment:

It is extremely important not to get dust or dirt inside of pressurised equipment as it could contain sparking or hot components. The equipment and the immediate surroundings should be cleaned first to avoid any risk of dust entering the enclosure when opened for Inspection. After the Inspection the equipment should be cleaned internally to remove any remaining dust inside before the power is restored. This will be discussed in the Exp pressurisation section.

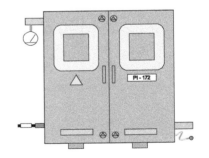

# Visual Inspection
## Extra Checks

Introduction:

Of course **Code 'E' numbers'** do not exist in the Standard. These are 'extra' checks that I consider should be done at this time and are on the bottom of the checklist. I have chosen a junction box as an example. **Engineers decision.**

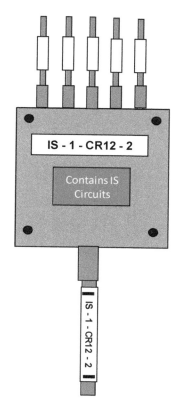

E1   If the junction box only has IS circuits within then it should have a blue label explaining 'contains IS circuits only.' **Does it have one?**

E2   If the junction box has more than one loop within it should be Certified. **Is the box Certified?**

E3   Are all circuit IDs present on all cables? With a **Visual** Inspection you cannot tell if they are correct as you just have an **Area Classification Drawing.**

E4   Are all the cover bolts present?

E5   Is the equipment ID available? Again with a **Visual** Inspection you can only see whether the ID is present, not if it is correct.

E6   Can you see any damage to the outside of the equipment?

E7   Can you read the Hazardous Area Data Plate explaining what **'Protection'** the box is? This is all you can check on a **Visual** Inspection.

# Close Inspection
## Extra Checks

There is only one 'extra' check on the Close Inspection which I can think of:

E8   Are the cover bolts tight?

# Detailed Inspection
## Extra Checks

(Not in the Standard)
Code E9 – E12

Introduction:

Of course, **Code 'E' Numbers'** do not exist in the Standard. These are 'extra' checks that I consider should be done at this time and are on the bottom of the checklist. I have chosen a junction box as an example.

The following 'Extra' points should be checked in my opinion

E9   Can you see any damage at all or anything missing inside the enclosure such as:

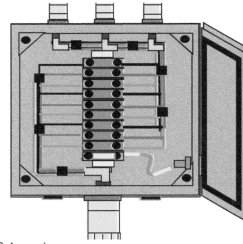

   a   Any damage to the connection block?

   b   Any Numbers missing off the block?

   c   Any damage to cable cores?

   d   Does the inner insulation of cables protrude into the box?

E10   Is there any copper showing from core end sleeves?

E11   Any signs of dust or water in the enclosure?

E12   Have all cable cores got core end sleeves fitted?  (Bootlace Crimps)

# Pressure Code Descriptions

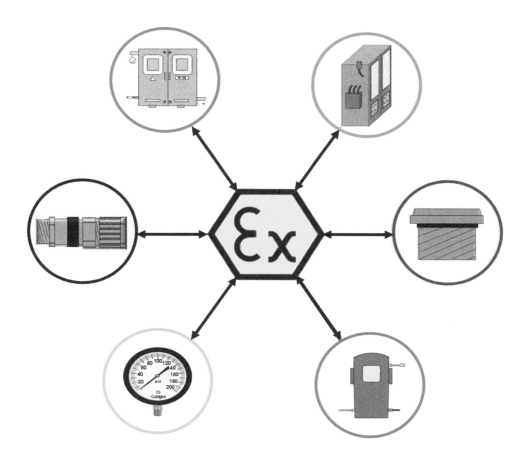

# Visual Inspection
## Is the Equipment Appropriate to the EPL/Zone Requirements of the Location?

## Code A1

### Introduction:

All that the Inspectors will have with them here is an **Area Classification Drawing.** This is a drawing with lines, circles or squares on it depicting the Zones. **No tools or access equipment.**

### Area Classification Drawing Markings:

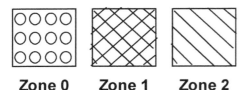

**Zone 0**　　**Zone 1**　　**Zone 2**

The **Gas Zone** markings will be shown on the **Area Classification Drawing** as shown to the left. These days Zone 2 will be the most common. Zone 0 usually inside of tops of vessels.

The **Dust Zone** markings will be shown on the **Area Classification Drawing** as shown to the right. Zone 20 usually inside tops of hoppers etc. Because of PPM Zone 21 areas will be rare.

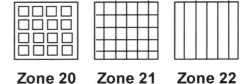

**Zone 20**　　**Zone 21**　　**Zone 22**

On this drawing there will be no Gas or Dust Groups, no temperature classifications or surface temperatures so although this information will be on the equipment there will be no reference on the drawing to match up.

### Protection/Equipment Protection Level (EPL):

What the Inspectors are looking for here is the 'Protection i.e. Exd, Exe, Ext, etc on the equipment to see if it is in the correct Zone **(see chart on next page).**

Manufacturers have started putting the EPL on the end of the markings, probably to get technicians and Inspectors used to seeing them. These are of course being introduced by the IEC, probably with a view to replacing the category. Gas Zone EPLs are Ga, Gb or Gc. As can be seen from the above diagram this **EPL is 'Gb'** i.e. suitable for **Gas Zone 1.** Dust EPLs will be Da, Db or Dc.

It must be noted that equipment suitable for Dust Zones will not automatically be suitable for Gas Zones and vice-versa. So if, say, the Inspector was to find **EPL Gb** in a **Dust Zone 21** this should be noted as a fault.

# Visual Inspection
## Is the Equipment Appropriate to the EPL/Zone Requirements of the Location?

## Code A1-1

| Protection: | Description: | Special Comments: | Zones: |
|---|---|---|---|
| Exd | Flameproof | Allows gas to enter | 1 & 2 |
| Exe | Increased safety | IP54 Minimum | 1 & 2 |
| Exec | Increased safety | Replaced **ExnA** | 2 ONLY |
| Exh | Mechanical | Combines Mechanical 'c' - 'b' - 'k' | Documentation |
| Exia | Intrinsic Safety | 2 Faults and remain safe. | 0, 1 & 2 |
| Exib | Intrinsic Safety | 1 Fault and remain safe. | 1 & 2 |
| Exic | Intrinsic Safety | Replaced **ExnL** | 2 ONLY |
| ExiD | Intrinsic Safety | Dust Atmospheres | Documentation |
| Exm | Encapsulated | Older type equipment | 1 & 2 |
| Exma | Encapsulated | Completely seals | 0, 1 & 2 |
| Exmb | Encapsulated | Completely seals | 1 & 2 |
| Exmc | Encapsulated | Completely seals | 2 ONLY |
| ExmD | Encapsulation | Dust Atmospheres | Documentation |
| ExN | Non Incendive | **Never Cenelec so no export.** | Withdrawn |
| ExnA | Reduced Risk | Non Sparking - Replaced by **Exec** | 2 ONLY |
| ExnC | Reduced Risk | Encapsulated | 2 ONLY |
| ExnC | Reduced Risk | Hermetically Sealed | 2 ONLY |
| ExnC | Reduced Risk | Enclosed Break | 2 ONLY |
| ExnR | Reduced Risk | Restricted Breathing | 2 ONLY |
| ExnL | Reduced Risk | Energy Limiting - Replaced by **Exic** | 2 ONLY |
| ExnZ | Reduced Risk | Pressurisation - Replaced by **Expz** | 2 ONLY |
| Exo | Oil Filled | Quenches any arcs or sparks | 1 & 2 |
| Exop (is) | Optical Radiation | Inherent Safety (is) | 0, 1 & 2 |
| Exop (sh) | Optical Radiation | Not Inherently Safe and Interlocked (sh) | Documentation |
| Exop (pr) | Optical Radiation | Protected Optical System (pr) | Documentation |
| Exq | Quartz / Powder Filled | Quenches any arcs or sparks | 1 & 2 |
| Expx | Pressurisation | Auto Shutdown | 1 & 2 |
| Expy | Pressurisation | No Shutdown | 1 & 2 |
| Expz | Pressurisation | No Shutdown - Replaced **ExnZ** | 2 ONLY |
| Expv | Pressurisation | Auto Shutdown | Documentation |
| ExpD | Pressurisation | Dust Atmospheres | 21 & 22 |
| Exs | Special Protection | **Older type equipment.** Never Cenelec | Withdrawn |
| Exsa | Special Protection | New Atex Certification | 0, 1 & 2 |
| Exsb | Special Protection | New Atex Certification | 1 & 2 |
| Exsc | Special Protection | New Atex Certification | 2 ONLY |
| ExtD | Protection by Enclosure | **Dust Atmospheres only** | Documentation |
| Exv | Ventilation | American Ex Standard | Documentation |

# Close Inspection
## Is the Equipment Group Correct?

### Code A2

#### Introduction:

What the Inspector is looking for here is the **'Group'**. This can take several forms e.g. **Surface Industry** or Mining, the **Gas Group** and the **Dust Group**. If the equipment is Atex, the Group will be in the form of Roman numerals after the Atex mark 'l' or 'll'. The predominant gases in each Gas Group are llA-Propane, llB-Ethylene & llC-Hydrogen. Dust Group equipment cannot, without the manufacturer's approval in writing, go into Gas Zones and vice-versa.

#### Mining Equipment:

If the equipment is for Mining, the **'Group'** is the **Roman l** after the Atex mark as shown in the diagram to the left. 'Ma' indicates the top section in mining.

#### Surface Industry Gas/Vapour Equipment:

As you can see in the diagram to the right, the **'Group'** is the **Roman ll** Surface Industry, after the Atex mark.

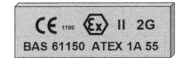

#### Gas Group ll:

The diagram on the left shows the **Gas Group** as just a **Roman ll**, not to be confused with the **ll** for Surface Industry.

If the **Gas Group** is shown as just a **Roman ll**, then the equipment will be suitable for any **Gas Group ll**. If the equipment is shown as **Gas Group llB + H2** then it is a **llB** piece of equipment, but also suitable for **Hydrogen** (not acetylene or carbon di-sulphide.) Find out the Gas Groups of all chemicals, gases & vapours on plant and ensure the equipment Gas Group is correct (which should be to the worst one**), llC being the worst.**

#### Surface Industry Dust Equipment:

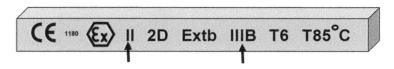

The **Group** will still be a **Roman ll** Surface Industry even though the dust groups are 'lll'. There is Dust Group lllA (Combustible Flyings i.e. jute, hemp, oakum, kapok etc), lllB (Non-conductive Dust i.e. wood, grain, starch, sugar etc.), or lllC (Conductive Dust i.e. coal, carbon, charcoal, magnesium etc.), **lllC being the worst.**

# Close Inspection
## Is the Equipment Temperature Classification Correct?

Introduction:

This of course cannot be completed until the **Close Inspection** as the information required would not be on the **Area Classification Drawing.** In this country we have six temperature classifications. The definition is: the surface of the equipment in contact with the gas, vapour or dust will not go over the temperature classification (T1 – T6) **under normal or specified fault conditions.** (It does not say it will never go over and the manufacturers will specify the faults.)

## Ignition Temperature:

| T1 | 450 |
| T2 | 300 |
| T3 | 200 |
| T4 | 135 |
| T5 | 100 |
| T6 | 85 |

Find out the ignition temperatures of all of the chemicals, gases, vapours and dusts on the plant and ensure that the temperature classifications of all the items of electrical equipment are below that value. In reality, Electrical Engineers will order T5 or T6 equipment to be on the safe side, even though this may be more expensive.

For example if you had a gas or vapour that had an ignition temperature of 235C. Looking at the UK chart left it would be no good using equipment with temperature classification T2 (300C) because if the equipment was to reach 300C under normal or specified fault conditions and there was a gas cloud with a 235C ignition temperature the 300C equipment would ignite it.

So the **Temperature Classification** of the equipment must be **under** the **Ignition Temperature** of the gases that it is in.

## Ambient Temperature:

The above definition applies in a normal **Ambient Temperature of -20C to +40C,** so if the equipment does not say anything else on the data plate then this is the **'Norm'.**

But what if the ambient temperature where the equipment is to be installed is higher or lower than the **'Norm'**? This situation needs different equipment so it's back to the manufacturers to obtain this new equipment. The data plate will show the higher or lower **Ambient Temperature** requested with **'Ta'** or **'Tamb'** stating what the new higher or lower Ambient Temperature is, **e.g. Ta = -50C to +55C.**

# Detailed Inspection
## Is the Equipment Circuit ID CORRECT?

Code A4

Introduction:

Circuit numbers are vital where isolations are concerned. On the **Visual** Inspection, Inspectors could only see if the numbers were **present**, but now on a **Detailed** Inspection they can check whether they are **correct** (extra documentation).

Some might argue that this is a **Close** Inspection, but possibly the standard thinking is that with a **Detailed** Inspection the numbers may have to be used in anger to isolate, so they can be checked at the same time.

How to decide what is required:

What do companies actually include in their numbers? If you look at the diagram on the left you will see that the number is P4 C3 meaning that this equipment is fed from distribution board P4 and circuit 3. Companies can make the number too long of course.

If the plant is new there can be situations where the cable numbers are incorrect as they may not yet have been used in anger for isolations, but if it is a well-established plant the chances are that the numbers have been used before so there is more chance of them being correct. Obviously the cable must have a clear number on either end of the cable.

### Are the Cabinet/Equipment and Circuit Numbers Present?

On some chemical plants like the one I worked on, pressurised cabinets and equipment is not that common, but they did exist. If we take a cabinet as an example, as you well know, these cabinets house equipment that might be quite hot or capacitive circuits that might require time to cool down or dissipate their charge.

The number might include the cabinet itself and the distribution board and circuit number where it is fed from. This being a **Detailed** Inspection, the Inspectors should have the documentation available to check if these numbers are correct.

As mentioned above, if the system is well established the chances are that the isolation has been done many times before and these numbers are correct. Other amenities such as pipework to and from the cabinet should also have numbers which will be on the documentation.

# Visual Inspection
## Is the Equipment Circuit ID AVAILABLE?

## Code A5

### Introduction:

Circuit numbers are vital where isolations are concerned. Now on this **Visual** Inspection the Inspectors can see if the numbers are **present**, but **not** if they are correct as they only have an **Area Classification Drawing.**

### How to decide what is required:

What do companies actually include in their numbers? If you look at the diagram on the left you will see that the number is P4 C3 meaning that this equipment is fed from distribution board P4 and circuit 3. You can make the number too long of course.

If the plant is new there can be situations where the cable numbers are incorrect as they may not yet have been used in anger for isolations, but if it is a well-established plant the chances are that the numbers have been used before so there is more chance of them being correct. Obviously the cable must have a clear number on either end of the cable.

### Are the Cabinet/Equipment and Circuit Numbers Present?

On some chemical plants like the one I worked on, pressurised cabinets and equipment is not that common, but they did exist. As you well know, these cabinets house equipment that might be quite hot or capacitive circuits that might require time to cool down or dissipate their charge. The number might include the cabinet itself and the distribution board and circuit number from where it is fed. This being a **Visual** Inspection, the Inspectors do not have the documentation available to check if these numbers are correct.

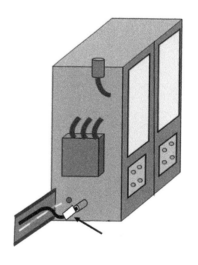

As mentioned above, if the system is well established the chances are that the isolation has been done many times before and these numbers are correct. Inspectors can check them on a **Detailed** Inspection.

Other amenities such as pipework to and from the cabinet should also have numbers.

# Visual Inspection
## Is there Any Damage to Glass to Metal Seals?

### Code A6

### Introduction:

Pressurisation allows the company to use uncertified electrical or instrument equipment in a hazardous area by manufacturers installing it in an enclosure and applying a positive pressure to keep out gases, vapours or dusts. Many of these enclosures have clear windows, enabling technicians to view the contents inside without opening the doors/covers. These doors/covers not only have to seal themselves against the environment, but the glass window also has to seal itself with the metal cabinet.

### Pressurised Cabinets:

If the seal between the glass and the metal is faulty on an enclosure like the one left, it is possible, but unlikely, that gas, vapour or dust will enter the enclosure because of the positive pressure. If there is a faulty seal between the glass and the metal then this may cause the pressure to drop slightly and nitrogen could escape here when purging.

The cabinet may have lamps fitted. Ensure that the glasses in these lamps are in good condition and unbroken. It is unlikely that the lamp units will be intrinsically safe, they may just be low voltage.

The glass must also be able to withstand the force of the pressure within the enclosure without its seals rupturing.

### Instruments:

Instruments similar to the example on the right also have a glass window where the display inside the equipment can be viewed by engineering and process technicians. This clear window will also have a seal to the metal. Inspectors should ensure that the seal is intact and that there is no damage to the window.

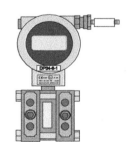

# Detailed Inspection
## Are there Any Unauthorised Modifications?

Code A7

Introduction:

When Inspectors looked at the outside of the installation on a **Visual** Inspection they only had an Area Classification Drawing and looked for external unauthorised modifications. Now the doors can be opened or the covers removed for an Inspection of the inside of the equipment. As this is now a **Detailed** Inspection more documentation is now available, so again look at the points below:

WARNING: ENSURE THAT CABINETS ARE NOT UNDER NITROGEN PURGE.

**Isolate the pressurisation air and follow any instructions or warning notices about time periods before opening the enclosure. There may be an interlock time. Inspectors check as per Checklist which might include:**

1   Any modifications not in the documentation?

2   Is all the equipment, i.e. terminal strips to Standard?

3   Does any equipment look to be added or removed?

4   Is there any tape around cables or pipes which might indicate that they are damaged?

5   Basically looking at the equipment, does there seem to be anything not quite right?

# Visual Inspection
## Is there Any Evidence of Unauthorised Modifications Outside the Equipment?

Code A8

Introduction:

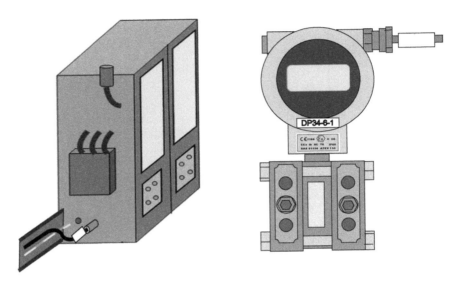

Look at the outside of the installation, remember at this stage on a **Visual** Inspection the Inspectors only have an **Area Classification Drawing**. Does it look as if there have been any unauthorised modifications? Inspectors check as per Checklist and faults may include:

1   Holes drilled in the equipment?

2   Any brackets added using the equipment screws?

3   Bolts used in one part that are completely different to all of the others in the equipment?

4   Is the gland correct? It does not have to be Certified in a cabinet so long as it fits the cable correctly and seals into the unit.

5   Is there any tape around cables or pipes which might indicate that they are damaged?

6   Basically looking at the equipment, does there seem to be anything not quite right?

# Detailed Inspection
## Are All Lamp Ratings & Pin Configurations Correct?

Code A9

Introduction:

When we talk about lighting & lamps we can split this subject into two sections. There are the coloured lamps that warn and indicate process & equipment conditions and there are the lamps that are required at night enabling people to see clearly and go about their business.

Then there is the size of the pressurised enclosure which might range from a small cabinet or instrument up to, say, an analyser building where people can actually go inside.

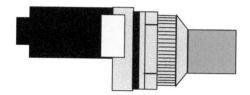

Let us take an instrument or cabinet with warning or indication lights on the front. These usually take the form of a transformer at the back leading down to a coloured glass as shown left. The supply may be 110V but the transformer allows a small LED lamp of 6V.

Check that the indicator lights are as per documentation.

Some cabinets have beacon type lighting along with an audible on the outside the unit to inform of interior problems (especially the Expx and Expy systems in Zone 1 areas). Ensure that these lights are up to Zone & Atex Standard. If there is any area lighting fixed to the cabinet, check that it is suitable for the Zone. If the pressurised enclosure is an analyser house in, say, a Zone 1 area, check that the fitting in the air lock is Certified to the Zone, Gas Group and Temperature Class. (Anyone entering the air lock could take gas in with them.) Also ensure that there is a self-contained light fitting inside the building to the correct Zone, Gas Group and Temperature Cass that is working correctly.

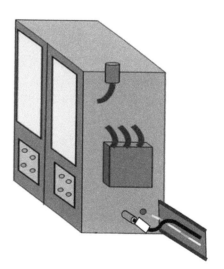

# Detailed Inspection
## Is the Type of Cable Appropriate?

### Code B1

#### Introduction:

When we ask if the type of cable is appropriate, we have to look at several points e.g. what is the sheath made of, what type of cable Protection, whether the cable is static or flexible etc. For instance, whether the cable is static or flexible might have a bearing on what type of gland is going to be used. What the sheath is made of could have a bearing on what type of chemicals will attack it if used on a chemical complex. Let us look at several different cables and how they might be used:

#### Flexible Cable (or Flex):

Exactly what it says, the cable is very flexible and can be used on portable appliances like hand-lamps, kettles etc where the cable does not require Protection like steel wire armour (SWA) or braid.

Another heat resistant flex is butyl rubber. Manufacturers of bulkhead light fittings usually use this type of flex to feed the lamp-holder from the connection block. The sheath is usually made of polyvinyl chloride (PVC) or cross-linked polyethylene (XLPE). The actual cable is very multi stranded to make it flexible. The gland for this type of cable would be a **compression gland** or stuffing gland as technicians refer to it, as there is no braid or SWA to earth. If this cable is installed on site then it must be protected by some means, such as trunking, conduit etc.

#### Steel Wire Armour Cable (SWA):

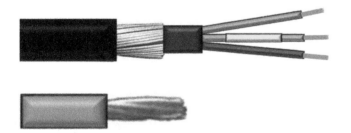

This cable is a bit more substantial and has the SWA as a cable Protector. Of course the SWA requires earthing and that is done by using a universal gland for SWA/braid. Not as flexible as the one above.

This SWA cable is usually round **extruded bedded,** which means the bedding (the inner PVC) is manufactured in one continuous length so it cocoons the conductors the full length of the cable. The sheath is usually polyvinyl chloride (PVC) or polyethylene (PE). The polyethylene is much harder and can withstand many chemicals which would affect the PVC. Cross-linked polyethylene (XLPE) surrounds the conductors. Inner insulation is Polyvinyl chloride (PVC) bedding. The cable can be obtained with aluminium wire armour (AWA) instead of steel wire armour (SWA). The steel wire armour should be earthed but not used as a sole earth (Company Policy).

# Detailed Inspection
## Is the Type of Cable Appropriate?

## Code B1-1

### Braided Cable:

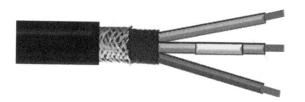

Braided cable has gradually become more popular than steel wire armour cable for wiring fixed systems in industry. The sheath can be ethylene propylene rubber (EPR).

There are two types of braided cable, the one used for fixed wiring systems similar to SWA, and the braided cable used for temporary supplies which is much more flexible. If the question was asked 'Why Braided Cable and not Steel Wire Armoured Cable' the answer might well be 'Braided Cable is lighter and more flexible'.

### Mineral Insulated Copper Covered (MICC):

MICC cable is a copper tube with conductors suspended in a white powder called magnesium oxide. Back in the 1960s and 1970s this was the only true flameproof cable as the magnesium oxide was the perfect filler since no explosion could get past it, therefore chemical plants tended to use it.

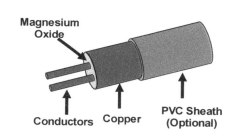

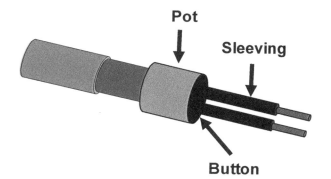

The magnesium oxide is very hygroscopic (water absorbent) so what is called a pot has to be put on the end filled with compound to make the cable end waterproof.

The compound and the way the pot is made off differs between an Exd gland and an Exe gland.

Back in the past I (the author) have made hundreds of MICC ends off and we found that the glands for this cable are Exd even though the pot may be made off Exe. See the MICC instruction manual for this type of cable.

# Visual Inspection
## Any Obvious Damage to Cables?

### Code B2

### Introduction:

When I mention here 'damage to cables' on a **Visual** Inspection, the inspectors would need to know in their remit as to how far up the line should they go? Inspectors cannot be expected to examine every metre of cable in great detail in, say, a 500 metre run, so a certain length from the equipment will be inspected in detail. The inspector may walk the entire run to see if there is any conspicuous catastrophic damage.

### Is there any local damage to Power Cables?

The cable trauma is, thankfully, not usually as devastating as the diagram above. Damaged power cables can be catastrophic to the plant as you can imagine. If the cable was damaged whilst isolated and the electrical technician replaces the fuses in, say, a lighting, motor or socket circuit then there would be a cloud of sparks where the damage was, making it imperative that any **Visual** Inspection makes every effort to pick this up.

Inspectors need to check to see if they can spot any local damage to the cables, cable tray or whatever cable management is being used, e.g. ladder rack, conduit, trunking etc. Damage may be found with steel wire armour showing, or even conductors in extreme cases as above:

### ON FINDING SOMETHING LIKE THIS DO NOT TOUCH IT.

If any catastrophic cable damage is found it must be reported to the control room and engineer **IMMEDIATELY,** not just noted down in some documentation. The circuit must be isolated as soon as possible, but in a correct, controlled manner. If you were to just pull the fuses you may send the plant into a crash shutdown, creating many other problems. These days the plant is most likely a Zone 2 area so not expecting gas to be present. If the cable is showing steel wire armour (SWA) on a vertical run it must be reported straight away as water can enter here and bypass the outer sheath seal on the gland.

# Visual Inspection
## Any Obvious Damage to Cables?

## Code B2-1

Is there any local damage to IS Cables?

The cable trauma is, thankfully, not usually as devastating as in the diagram above. Intrinsically safe cables (Light Blue) showing damage should again be reported immediately. Although probably not dangerous as far as sparking and ignition is concerned, this type of damage could cause instruments to malfunction, the plant to go into shutdown etc. Can Inspectors see any damage to the cables, cable tray or whatever cable management is being used e.g. ladder rack, conduit, trunking etc? They may find damage with steel wire armour showing or even conductors in extreme cases as above. On finding something like this do not touch it. On finding any catastrophic cable damage this must be reported to the control room and engineer **IMMEDIATELY,** not just noted down in some documentation.

Is there any damage to cable at the Gland?

The situation on the diagram to the right might be more common where the steel wire armour is showing at the gland. This might be because someone has not made the gland off properly in the first place. If the cable is showing steel wire armour (SWA) or braid on a vertical run it must be reported straight away as water can enter here and bypass the outer sheath seal on the gland.

Example of Water entering Cable Sheath?

I came across a situation where one of our motors was filling with water in its terminal block and no-one could find out where the water was coming from. The cable had a polyethylene sheath, which is extremely hard and where it came off the cable ladder rack many feet above the motor there was a huge split in the sheath as in the diagram on the left.

So water was entering here and travelling many feet down to the motor, bypassing the gland and into the terminal box. We taped up the flaw very thoroughly with special tape, solving both the mystery and the fault. I thought I would mention this, although on a **Visual** Inspection there would be no access equipment.

# Visual Inspection
## Is the Earth Bond Connected to Earth &
## Is the Earth Wire Sufficient Cross Sectional Area?

## Code B3

### Introduction:

What is the difference between earthing and bonding? The answer may be, not a great deal. Equipment is earthed to give a path back to the star point of the electrical distribution transformer so that protective devices such as fuses or earth leakage circuit breakers can operate in the event of a fault in the equipment which may otherwise leave it live. This is usually achieved by a third green & yellow core in the feeder cable. Bonding is mainly to stop two parts of a metal system from becoming different in potential (for instance equipment and the metal cable tray), and is usually achieved by an external green & yellow wire.

### Is the Earth Bond present?

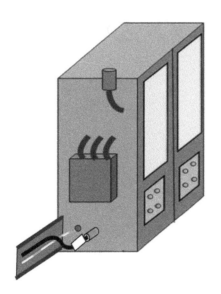

If the equipment is 'end of line', meaning that the circuit stops here, such as a cabinet like the one shown left, technicians would usually fit an earth tag connected to earth if the equipment was not metal. Manufacturers may give advice. Check company policy here.

As Inspectors can see in the diagram (left), the cable enters at the bottom and should have a serrated washer and locknut on the inside if it is a clearance hole. The cable should be 3 core so that the SWA is not being used as the sole earth.

This is a **Visual** Inspection so doors would be closed, but a point is that if there are electrics on the door then the door should be bonded which would be picked up on a **Detailed** Inspection.

### Is the size of the Earth Bonding Conductors 4mm?

The earth bond wire should be a minimum of 4mm². This cable must be capable of carrying any earth currents and, at the same time, offering least resistance.

# Detailed Inspection
## Earth Loop Impedance of a TN System
## (Terre Neutral Separate)

## Code B4

### Introduction:

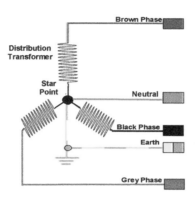

### Why do an Earth Loop Impedance Test?

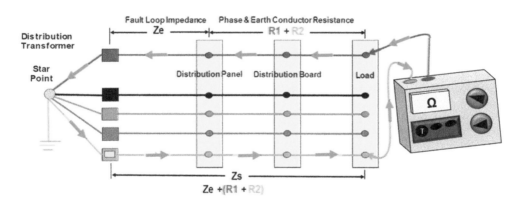

What are Inspectors doing with an earth loop impedance test? Well you could say that by utilising the live and earth conductor, we are driving electrons through the star point of the distribution transformer as shown by the red arrows on the above diagram, and obtaining a reading in $\Omega$ of that circuit. Quite a high current is involved, around 20 amps. Tests are from the furthest point from the distribution transformer if possible.

We have to ensure that if a fault did occur in an installation, the current that flows would be enough to operate the protection systems such as circuit breakers or fuses and it must be ensured that these devices will operate in a certain time. Just imagine what could happen to hazardous area equipment if the protection devices did not operate fast enough:- **LIVE EQUIPMENT/HEAT/SPARKING.**

**Readings as per 18th Edition IEE Regulations (BS 7671)**

# Detailed Inspection

## Earth Resistance of an IT System? (Isolated Terre)
## Earth Resistance of TT System? (Terre Terre)

## Code B4-1

### Introduction:

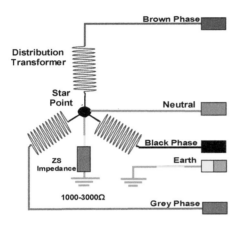

Here in the **IT system** (Isolated Terre) a high impedance (Zs) is inserted into the distribution transformer star point earth, sometimes no impedance at all so that the star point is not earthed.

The idea is that in the case of an earth fault on site it would cut down the high amperage fault currents if the star point had no earth or a high impedance one. **TT systems** (Terre, Terre) can be tested the same.

### Testing the Earth Rod:

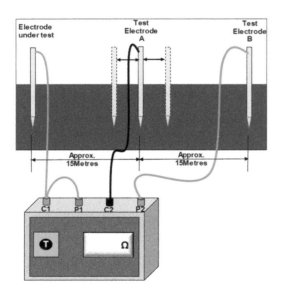

In the diagram to the left you can see the earth rod under test on the left hand side. Two test rods are then put into the ground, namely rod A & rod B. Rod B is put into the ground 25 – 50 metres away from the rod under test. Rod A is put into the ground about midway between rod B and the rod under test.

These Rods are connected to the earth tester (Megger) as shown in the diagram, and a reading taken in Ω. Rod A is moved both nearer and further away from the rod under test as shown in the diagram, and at each position a reading is taken.

The most constant reading is the rod resistance that we are after. Many years ago in my day as a young electrical technician, this would be done with a bridge Megger as in Wheatstone bridge. The principle here is a current & potential loop hence the C & P on the tester terminals, explained in my book 'Hazardous Areas for Technicians'.

# Detailed Inspection
## Automatic Electrical Protective Devices are Set Correctly?

## Code B5

### Introduction:

When it comes to automatic protective devices, electric motor overload units spring immediately to mind. These units can be set to, in most cases, **manual reset** and in other cases, **automatic reset**. The time delay to trip can be adjusted within a small band depending upon the full load current of the motor that they are protecting. A set of overloads are chosen to ensure that the full load current of the motor is around the middle i.e. If the motor full load current is 4 amps a set of thermal elements might range from 2 amps to 6 amps. It is possible to obtain a px or a py pressurised motor.

### NO Automatic Reset:

On a normal three phase, squirrel cage, induction motor the overloads would be set to '**Manual Reset**' meaning that, should the motor go into overload, the unit will trip, a small button will pop out and this must be then pushed in by hand to reset it. In the past if the motor was in a sump or well it might be that, for this purpose, the manufacturers would build an **Automatic Reset** into the overload unit. The motor could have thermistor units in the windings in case the motor got too hot.

**Automatic Reset** is, in my opinion, a very dangerous facility as if the motor did have any real problem with overload, the overload unit would just keep on resetting after a cool down period, and if the motor was also set to Automatic Start, it would just keep starting and heat up. These days, especially on **Exe motors, the overload unit must NOT be set to Automatic Reset.** Automatic Reset may be a good idea on equipment such as fridges.

### Current Transformer Overloads:

If the motor is quite large to save very large currents passing through the overload unit hence increasing its physical size enormously it is possible for the manufacturers to build starters where the large currents pass through current transformers which in turn allows for physically smaller overload units.

Injecting using the CTs is called **Primary Injection** and requires a much larger injection set to obtain the current required in the CT primary ratio. **Secondary Injection** is preferred because of the smaller current injection.

**See next section B9 for connections and example injection currents:**

# Detailed Inspection
## Do Automatic Protective Devices Operate Within Permitted Limits?

### Settings:

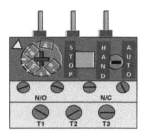

Overloads are set to the full load current of the motor obtained from the motor data plate.

This should tie in with the red load line or, as it is nicknamed, the 'Bloodline' on the ammeter.

### Secondary Injecting the Overloads:

Inject the overloads as per manufacturer's Instructions this should be something like the following:

Connect all of the overloads in series as per the diagram on the right ensuring that the current flows in the same direction through each phase overload (or see manufacturer's documentation). Suggested injections: From cold, inject **150%** of full load current and see how long they take to trip (minutes). From hot, as soon as the overload will reset inject **300%** of full load current (seconds). As soon as the overloads will reset, inject **108%** of full load current (should not trip in 8 minutes).

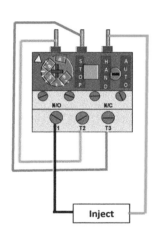

### Single Phasing Injection:

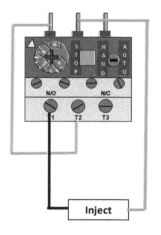

Inject the overloads as per manufacturer's instructions which should be something like the following:

Disconnect one phase of the overload unit as per the diagram on the left (i.e. the purple cable) and connect the Injection set as per the red and black cables. Now inject **105%** of the full load current and ensure that the overloads trip in a manufacturers' given time. This test is to ensure that the motor is protected from single phasing because, say, of one fuse blowing because, of course if this happened, the current would rise on the other phases.

# Detailed Inspection
## Is Protective Gas Inlet Temperature Below Maximum Specified?

Code B7

Introduction:

The inlet temperature of the pressurisation gas will be quoted by the manufacturers. It must be noted that, although you might think that the pressurisation gas is quite cool, it must not be relied on alone to cool any components within the equipment which, during their operation, are likely to get hot. It is possible to obtain air conditioning units to seal into the enclosure that specifically provide cooling for the internal components.

The Pressurisation Protective Gas:

The pressurisation protective gas, which is usually air, (being the cheapest and least hazardous to exhaust), will be at a pressure and temperature recommended by the manufacturers. The reading recommended must not stray beyond tolerance and a manufacturer's data sheet should always be obtained before any Inspection.

The Compressor:

The pressurised equipment compressor is usually located in the non-hazardous area very close to the clean air intake, and because of this location, the motor and compressor do not need to be Certified. Pressurisation for equipment may be slightly higher than that for buildings which will be above atmospheric pressure. The pressure must not be too high or there may be a risk of damage to seals.

The protective gas has of course to be at the correct pressure and temperature, which again is usually below the maximum specified by the manufacturers. The main concern with pressurisation is to keep a contaminated outside atmosphere out, rather than to keep the protective gas in.

# Visual Inspection
## Are Ducts, Pipes and Enclosures in Good Condition?

## Code B8

### Introduction:

Pressurisation allows equipment with a lesser Certification to be used in a hazardous area by using pressurised air to separate it from any hazardous gas or vapour by causing what is called a positive pressure.

The pressurised equipment compressor is usually in the non-hazardous area very close to the clean air intake and because of this, the motor and compressor do not need to be Certified. Pressurisation for equipment may be slightly higher than that for buildings which is 0.5mBar above atmospheric pressure. The pressure must not be too high or there may be a risk of damage to seals.

The protective gas has of course to be at the correct pressure and temperature, which again is usually below the maximum specified by the manufacturers. The main concern with pressurisation is to keep a contaminated outside atmosphere out, rather than to keep the protective gas in.

### Doors:

Some pressurisation cabinets just have the one door. Some cabinets have two doors, but they must open and close individually and not overlap (so you must have to open one door before you can open the other). It is difficult to obtain a good seal on overlapping doors. The doors are usually opened by special 'key spanners' and do not simply have two handles.

### The Enclosure:

Look at the enclosure outside, is there any corrosion or damage that might compromise the certification? Is the paintwork/finish satisfactory?

### Ducts & Pipes: (Visual Inspection)

All ducts & pipes must be checked for corrosion and leakage which should be reported if suspected. Pressure devices may be opened and checked only on the **Detailed** Inspection.

# Visual Inspection
## Is the Protective Gas Substantially Free from Contaminates?

## Introduction:

Pressurisation allows equipment with a lesser Certification to be used in a hazardous area by using pressurised air to separate it from any hazardous gas or vapour by causing what is called a positive pressure.

The pressurised equipment compressor is usually in the non-hazardous area very close to the clean air intake and because of this, the motor and compressor do not need to be Certified. Pressurisation for equipment is slightly higher than that for buildings which is 0.5mbar above atmospheric pressure. The pressure must not be too high or there is a risk of damage to seals.

The protective gas has of course to be at the correct pressure and temperature, which again is usually below the maximum specified by the manufacturers.

## Is the Protective Gas Free from Contaminates:

This may seem quite a question for a **Visual** Inspection, but all that is required here is:

1    Have there been any changes to the Zone since the equipment was installed?

2    Have any new plants been built since the commissioning of the system under inspection?

3    Is the protection air drawn in fact from the non-hazardous area?

## Air Quality:

Usually on a plant there are several types of air compressor. **Breathing Air,** which is clean and dry, used for what it says, breathing masks. **Maintenance Air,** where air tools are used, this can be very damp and is sometimes not so clean with very basic filters so could get contaminated. **Instrument Air,** which must be clean and dry and feeds delicate instruments, and **Pressurisation Air,** which is on the same sort of level as Instrument Air and must also be at the correct temperature. In other words, very high quality air is what we are interested in at this time.

# Visual Inspection
## Is the Gas Pressure/Flow Adequate?

## Code B10

### Introduction:

Pressurisation allows equipment with a lesser Certification to be used in a Hazardous Area by using pressurised air to separate it from any Hazardous Gas or Vapour. The pressurised equipment compressor is usually in the Non-Hazardous Area very close to the clean air intake and because of this, the motor and compressor do not need to be Certified. Pressurisation for equipment is higher than that for buildings. The protective gas has of course to be at the correct pressure and temperature, which again is usually below the maximum specified by the manufacturers.

### Protective Gas Flow/Pressure:

Look at the pressure gauge and check if the air flow/ air pressure is at the specified limits. The gauge should be marked and the pressure may be on the data plate or in the equipment documentation.

**THE EXACT FIGURES RECOMMENDED BY THE MANUFACTURER ARE TO IEC STANDARD AND MUST BE CHECKED FOR THE PRESENT AUTHENTICITY WITH THE STANDARD.**

### Over-pressurisation:

There are usually over-pressure valves & relays built into the system to guard against over-pressure. These must be tested to manufacturer's specification figures, although sometimes they may not be inspected until the **Detailed** Inspection.

### Exhaust:

Ideally the air is exhausted into a non-hazardous area. Failing this, the outlet must have a spark arrestor fitted local to the equipment. Which of these is installed?

# Detailed Inspection
## Do Pressure and /or Flow Indicators, Alarms & Interlocks Function Correctly?

## Code B11

### Introduction:

It is vital that loss of pressurisation activates audible/visual alarms to let the instrument technicians know there is a problem. A chemical factory, for instance, might prefer to have visual alarms as well as audible alarms, as plant technicians may be wearing hearing protection. These alarm activators must be tested on **Detailed** Inspections to ensure they work. Any interlock systems can also be function checked at this time.

### Purge & Pressurisation Principle & Gases:

A purge of, usually Nitrogen, is completed prior to pressurisation to rid the enclosure of any unwanted hazardous gas or vapour. There may be several purges **(possibly up to 5 air changes)** at one time (usually explained on the equipment label) especially if the equipment has been shut down for any length of time. The purge gas, as mentioned, is usually Nitrogen and somewhere there will be either several smaller bottles or one large storage tank. The pressurisation gas is usually air (from a clean air intake). So no equipment inside of the enclosure may be energised until after the purge cycle(s) are completed.

### Different Systems:

Expx (Zone 1 & 2) **Automatic** Shutdown should pressurisation be lost.

Expy (Zone 1 & 2) No immediate Shutdown should purge be lost.

Expy (Zone 2 **Only.**) No immediate Shutdown should purge be lost.

### Over-pressurisation:

Over-pressure relays can be tested on the **Detailed** Inspection to ensure that they will activate if the pressurisation pressure exceeded the maximum limits.

# Detailed Inspection
## Is the Condition of Particle and Spark Barriers of Ducts for Exhausting the Gas in Hazardous Area Satisfactory?

## Code B12

### Introduction:

It is not always possible to vent the exhausted air into a non-hazardous area, sometimes it has to be vented into the hazardous area. As mentioned in earlier sections, although the air might be quite cool (to manufacturer's temperatures) when it enters the equipment, any hot component inside will heat it up slightly. Also what has to be catered for is a fault condition inside of the equipment where there might actually be a small explosion. Hot gases and particles cannot be allowed to escape the enclosure assisted by the pressurisation system.

### Exhaust Gas:

What the Inspector is looking for here is whether the pressurisation gas exhausts into a non-hazardous area. If not, is there a spark arrestor, (also called spark traps), fitted and is it, together with the pipework associated with it, in good order? Any tests on this unit must be completed as per manufacturer's instructions.

### Spark Arrestor Example:

As well as the above, a good example of a spark arrestor which, by the way, will be larger, is on a motorbike silencer. The hot gases and sparks from the engine are decreased.

### How does a Standard Spark Arrestor Work?

Many have a static fin system inside which causes the hot gases to 'swirl' towards the walls of the spark arrestor where the turbulence causes them to lose their heat. Some cause carbon to be removed from the gas and may have to be cleaned at intervals to remove these deposits.

### Electrostatic Precipitator (ESP):

It all depends which type of device is used. An electrostatic precipitator uses an induced electrostatic charge to remove particles and hot gases. This is a simple, efficient method.

# Detailed Inspection
## Are All Conditions Of Use Complied With?
## (Items with an 'X' after the Atex/BASEEFA Number)

## Code B13

### Introduction:

This section refers to a scenario where the equipment has an **'X'** after the Atex number. The **'X'** refers to **'Special Conditions of Use'** which will also be stated on the Certificate which comes with the equipment. The Inspector must look at the Certificate, note the Special Conditions and see that they are being complied with.

### Where is the 'X'?

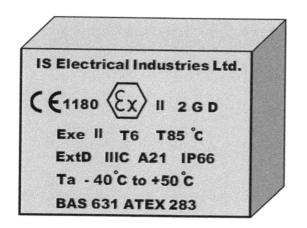

As you can see on the diagram to the left the 'X' follows the Atex or Baseefa number. What Inspectors are looking for now is the Certification that came with the equipment so that they can see what the Special Conditions of Use are and then see that they are being complied with.

It is very important that the Installer is shown the Certification so that they are aware of what Special Conditions they have to abide by when installing the equipment.

### Three examples of Special Conditions:

| | | |
|---|---|---|
| Enclosures assigned an impact level of 4J or 4NM and must be installed only in areas of low mechanical danger | Precautions must be taken to ensure that the thickness of dust layer on the terminal box will not exceed 5mm | Wiring within the enclosure must not be grouped or bunched to prevent hot spots forming |

# Visual Inspection
# Is the Equipment Protected Against the Weather?

## Introduction:

We cannot put our hazardous area electrical and instrument equipment inside a bubble and protect it from all weathers. The equipment ingress protection (IP) will usually protect it against solids and liquid ingress. What about direct sunlight and, at the other end of the scale, extreme cold? Direct sunlight can obviously cause the equipment to heat up.

Older equipment has a temperature classification which usually requires the normal ambient temperature to be between **-20C to +40C.** More modern equipment has higher and lower ambient temperatures, but might be worth investigation. Another point to consider is that at very low temperatures and very high temperatures, certain electrical insulation materials will degrade and, to some extent, lose their insulating property. If the ambient temperatures are known to be higher or lower than the 'norm', equipment with Ta or Tamb can be obtained with a higher or lower ambient temperature.

## Very Low Temperatures:

The inspector(s) should be concerned as to whether in very low temperatures the hazardous area equipment can maintain the 'Protection' that make it a Certified item i.e. Exd, Exe, Exi, by way of the material used to construct the equipment, the norm being -20C to +40C. So:

1    **Can the equipment function efficiently at very low temperatures, for instance if the temperature was to go below the norm of -20C?**

2    **Could the surrounding atmosphere gas/vapour change its properties at very low temperatures?**

3    **Can the equipment function efficiently at very high temperatures, for instance if the temperature was to go over the norm of +40C?**

## High Temperatures (e.g. Direct Sunlight):

Heat can cause electrical insulation degeneration. Cables in conduits which are in direct sunlight can be subjected to very high temperatures. PVC conduits & pipes can become more brittle when subjected to direct sunlight due to ultra violet (UV) light on the PVC. Heat can build up outside and inside enclosures, especially if they are made of metal. Manufacturers may give guidance on ambient temperatures.

# Visual Inspection
## Is the Equipment Protected Against Corrosion?

## Introduction:

Corrosion is usually caused by some kind of chemical reaction. In many cases it is because the equipment is the **'Anode'**. By using **'Cathodic Protection'** we can cause the equipment to be a **'Cathode'** instead. This can slow the corrosion down enormously. Below we talk about looking for corrosion on Inspections and what signs the Inspectors might be looking for.

## Inspecting Equipment for Corrosion:

Inspectors carry out the inspection, if there is any corrosion they must Risk Assess:

1    **What is the corrosion and what equipment is it on?**

2    **Is the corrosion minimal and requires no action?**

3    **Is the corrosion acceptable at the time of Inspection but a Risk Assessment determines it will get worse in the near future?**

4    **Is the corrosion great at the time of the inspection, requiring maintenance work straight away?**

## What is Corrosion?

I always look at corrosion as being released energy that was trapped into, say, steel at its liquid state and the metal is now returning to a more stable state, e.g. Oxides, Sulphides or Hydroxides. If we look at any rusty structure that is made of steel, for instance a rusty pipeline without any cathodic protection, it is not all rust (Ferrous Oxide), there are some good areas here and there. Where those good and bad areas meet, a cell is formed between the good parts (Cathodes) and the corroding parts (Anodes), the electrolyte (Water) provided by moist air or soil.

## Cathodic Protection:

We can use cathodic protection on a pipeline to slow this process down, explained in the book **'Hazardous Areas for Technicians' ISBN: 978-1-912014-95-8.**

| Cathodic | Platinum |
|---|---|
| ↑ | Gold |
| | Titanium |
| | Silver |
| | Tin |
| | Lead |
| | Mild steel |
| ↓ | Aluminium |
| Anodic | Zinc |

If we take cable tray, the galvanising is a type of zinc coating cathodic protection. If you look at the cathodic table left you will notice that zinc is more **'Anodic'** than mild steel hence the zinc forces the mild steel to be a **'Cathode'**. Corrosion will always go for the anode i.e. the Zinc. If you look at Gold you will see that it is near the top of the table i.e. very cathodic, which is why you can dig up gold coins out of the ground and there is hardly any corrosion at all.

# Visual inspection
## Is the Equipment Protected Against Vibration?

## Code C1-2

### Introduction:

There are many things that can cause vibration in chemical factories and on platforms. In the electrical world, electric machines spring straight to mind e.g. motors, generators etc. In the mechanical world there are compressors, pumps etc. In the process world there are the plant processes themselves. Vibration causes loose threads, bad connections and cracking where there is a metal seam.

### Vibration effects on several instruments:

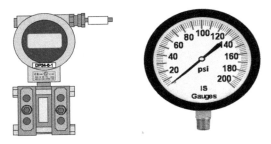

There are several instruments that vibration may have a very adverse effect upon, let us look at a few:

1   Pressure Sensors: vibration can severely disturb the measurement signal as it can be transmitted to the output signal.

2   Transmitters: very susceptible to prolonged vibration. Most manufacturers will recommend that they be mounted, if possible, in areas with the lowest vibration.

3   Load Cells: are susceptible to vibration and again, this can be transferred to the output and give inaccurate readings.

4   Pressure Gauges: Pressure gauges, by the mere fact of their mechanisms, are very susceptible to vibration. Oil filled gauges are less likely to fail due to vibration than a standard pressure gauge.

Where possible, instruments should be mounted in areas of least vibration, however, saying that is easy, carrying it out on plant may be extremely difficult. The location of the instrument depends upon the process. It may be an idea to use some vibration monitoring equipment in the event of several failures, and obtain some evidence of exactly how much vibration there is.

# Visual Inspection
## Is There Accumulation of Dust or Dirt?

### Code C2

### Introduction:

If dust or dirt is allowed to build up on the equipment then this could have a detrimental effect on the temperature of the equipment. On a **Visual** Inspection you are just looking to see if there is a build-up of dust or dirt on the equipment.

### Temperature Classification in General:

| T1 | 450 |
|----|-----|
| T2 | 300 |
| T3 | 200 |
| T4 | 135 |
| T5 | 100 |
| T6 | 85  |

The table to the left shows the six temperature classes in the UK. Equipment should not go over these temperatures under **NORMAL AND SPECIFIED FAULT CONDITIONS**. The ambient temperature for these temperature classes has a norm of **-20C to +40C**

If the Ambient Temperature is higher or lower, then alternative equipment must be obtained, in which case the data plate will state **Ta** or **Tamb** and the new higher and/or lower ambient temperature will be stated.

### How hot are these above Temperatures?

These temperatures are taken for granted, but do you realise how hot the equipment would be if it reached 450C as in T1? Look at the effect that temperature has on certain materials, some of which are used in the electrical world.

In the table on the right you may be able to get some idea of what can really happen if things start to heat up due to fault conditions.

| 455 | PVC Ignites |
|-----|-------------|
| 327 | Lead Melts |
| 232 | Tin Melts |
| 135 | Polyethylene Melts |
| 100 | Water Boils |
| 85  | Burn Hands |

WE DO NOT WANT ANY CONDITIONS WHICH MAY CAUSE THE EQUIPMENT TEMPERATURE TO RISE AND BEING COVERED IN DIRT AND DEBRIS IS ONE OF THEM.

# Visual Inspection
## Is There Accumulation of Dust or Dirt?

## Code C2-1

### Dust Ignition Temperatures:

If we just work on a golden rule that 'clouds explode and layers burn', the table (right) shows the ignition temperatures of various dusts with the different ignition temperatures of their clouds and layers. Our equipment temperature must not get anywhere near to these temperatures or there could be an explosion (Cloud) or fire (Layer). Excessive dust or dirt beyond 5mm could smother the equipment and cause it to heat up so good housekeeping is vital.

| Dust | Ignition Temperatures °C | |
| --- | --- | --- |
| | Cloud | Layer |
| Cellulose | 520 | 410 |
| Flour | 510 | 300 |
| Grain | 510 | 300 |
| Sugar Dust | 490 | 460 |
| Tea Dust | 490 | 340 |
| Starch | 460 | 435 |
| Lignite | 390 | 225 |

### Dust & Dirt on IS equipment:

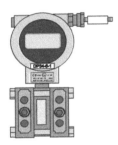

Dust & Dirt on IS Instruments will not usually cause heat as there is no real power there. What it may cause, especially in the case of a transmitter, is for the instrument to malfunction. For example sometimes an instrument takes its reference from atmospheric pressure. If any debris clogged up the reference part of the instrument it would malfunction. Dust & dirt on metal equipment can also cause corrosion so good housekeeping is essential.

### Dust & Dirt on Exp Pressurised Equipment:

It is extremely important not to get dust or dirt inside pressurised equipment as it could contain sparking or hot components. The equipment and the immediate surroundings should be cleaned first to avoid any risk of dust entering the enclosure when opened for Inspection. After the Inspection the equipment should be cleaned internally to remove any remaining dust inside before the power is restored.

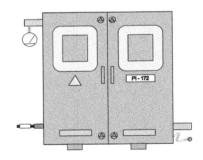

# Extra Points for Equipment Protection Exp:

The official IEC60079 Standard Codes go from **A1 – A31, B1 – B23** and **C1 – C3.** This is where the codes end.

On the Checklists you will see items that I consider require inspecting at this point, but are not listed in the Standard.

I have studied the codes and have come to the conclusion that there are some points that require an Inspection that are not in the codes. I have listed these below and described the new points on the following pages. I have called them E1 - …   (E meaning **'Extra to the Standard Codes'.** You will see what I mean as you look though the list:

E1   **(Visual)** Is there any damage to the outside of the Equipment?

E2   **(Visual)** Is the Equipment ID available?

E3   **(Visual)** Are all Stoppers present and no holes in the Equipment?

E4   **(Visual)** Do all Glands look sealed into the Equipment?

E5   **(Close)** Is the Equipment ID correct?

E6   **(Close)** Is the Earth Bonding tight?

E7   **(Close)** Are all of the Glands tight?

E8   **(Close)** Are all the Stoppers tight?

E9   **(Detailed)** Are all Cables satisfactory within the Enclosure?

E10   **(Detailed)** Are the Gland Internals correct?

E11   **(Detailed)** Is there any evidence of Water in the Enclosure?

E12   **(Detailed)** Is there any Evidence of Dust in the Enclosure?

E13   **(Detailed)** Is the Enclosure Seal satisfactory?

E14   **(Detailed)** is there any damage to Components within the Enclosure?

# Mechanical Descriptions

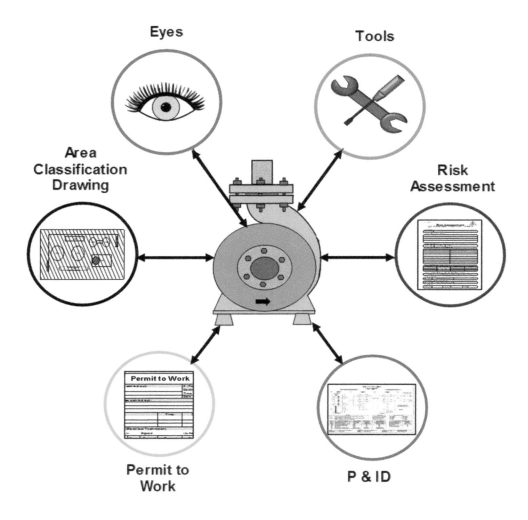

# Visual Inspection
## Flanges

## M1

### Introduction:

In the mechanical world there are only two grades of Inspection, namely **Visual** and **Detailed.** There is no **Close** Inspection, so Mechanical Inspectors do not quite come under the same Inspection criteria as the Electrical Inspectors, for example, they can discover loose nuts/bolts on a **Visual** Inspection. Many of the pumps, for instance, that are inspected at the moment will not be Atex, therefore a Risk Assessment should be completed by the Mechanical Engineer. So what **Visual** checks would a Mechanical Inspector carry out?

### Is there a Gasket between the Flanges?

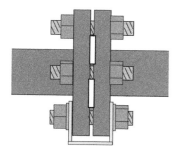

If there is a gasket it should be able to be seen. If the product were to pass down the pipeline with no gasket, the flange would definitely leak. That could be anything from a drip to a high-pressure spray. If the gasket is made of incorrect material, the product could leak in the near future as it destroys the material of the gasket, but this would not be found until a **Detailed** Inspection was undertaken.

### Is the Flange Bonding Link intact?

If the link is not put in, with very fast flowing liquid (especially an insulator liquid), or if it flows through a fine filter, a static charge can build up on the pipework. Nuts & bolts may not be enough for a good connection across the flange due to corrosion, paint, grease, non-conductive surface etc so an extra flange bond may be required. Just a point: remember that the gasket itself is an insulator.

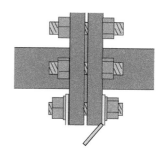

A Risk Assessment is carried out by the engineer to see if this 'Bonding Link' is required. If the product in the pipe is a liquid or dust that will carry static charge e.g. toluene, heptane, carbon-disulphide etc, this may influence the decision.

### Are any of the Nuts Loose?

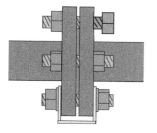

Loose nuts on a flange as left could, of course, allow product to escape so this is quite dangerous and needs to be discovered.

An important point to remember is that if a mechanical technician at BP found a loose nut, they must not attempt to tighten it whilst the pipeline was under pressure in case the through bolt sheared.

# Visual Inspection
## Flanges

## M1-1

### Are any of the Stud Bolts too long?

There are certain codes that stipulate how long the stud bolt should be, based on how many threads protrude through the nut. Engineer/company policy may decide this but usually the norm is that at least one thread pitch should protrude.

I have never come across a Standard that states that the stud bolts can be too long. Cost and waste of material may be the decider here. Several of those shown in the diagram (right) would be too long in this statement.

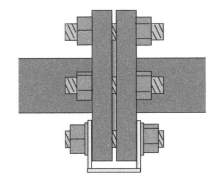

### Can the Stud Bolts be too short?

Most definitely the stud bolt should pass through the nut as above and have, say, one thread pitch protruding. If the stud bolt is not long enough it would stop one or two thread pitches short of the end of the nut (left). In a stress situation this may not be enough and the bolt may shear its end few threads.

### Can the Nuts be wrong way round?

Nuts are usually chamfered both sides (right) so that they are universal, but some do have a distinct flat side which must go to towards the flange, and some are what is called, flat 'washer faced' (left) which again must go towards the flange.

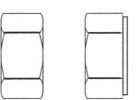

# Visual Inspection
## Valves

### Introduction:

In the mechanical world there are only two grades of Inspections, namely **Visual** and **Detailed.** There is no **Close** Inspection, so Mechanical Inspectors do not quite come under the same Inspection criteria as the Electrical Inspectors, for example, they can discover loose nuts/bolts on a **Visual** Inspection. Many of the pumps, for instance, that are inspected at the moment will not be Atex, therefore a Risk Assessment should be completed by the mechanical engineer. So what **Visual** checks would a Mechanical Inspector carry out?

### Valve Flanges:

The valve will have flanges, stud bolts & nuts which are dealt with in the previous section (M1).

### Is the Chemical Flow the same as the indicator?

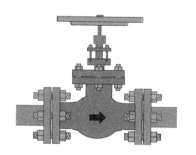

Does it matter if the valve is wrong way round and the direction arrow is against the flow? Let us just look at a globe valve. A globe valve is more for regulating the flow whereas a gate valve stops the flow. If the globe valve is installed the right way round then the pressure side is under the disc protecting the packing from pressure when the valve is turned off.

If the valve is installed incorrectly, the pressure will be on the top side thus reducing the ability to control the valve at the near closed position. There will be more chance of the valve leaking due to pressure on the packing, and maintenance on the valve whilst in place would be impossible. So the answer to the question is:

YES, IT DOES MATTER WHICH WAY ROUND THIS VALVE IS INSTALLED.

Gate valves, on the other hand, are generally used in circumstances where the flow is either on or off so they should be fully open or fully closed. They should not be used for flow control as damage can occur. They are called 'Gate' valves because of the wedge shaped plug inside resembling a gate. In most cases it does not matter which way round the valve is fitted, although they should not really be installed upside down or on a vertical line because of silt deposit on the gate guideways.

SO NO, IT DOES NOT USUALLY MATTER WHICH WAY ROUND THIS VALVE IS FITTED (CHECK MANUFACTURER'S INSTRUCTIONS).

# Visual Inspection
## Pumps

### Introduction:

In the mechanical world there are only two grades of Inspections, namely **Visual** and **Detailed.** There is no **Close** Inspection so Mechanical Inspectors do not quite come under the same Inspection criteria as the Electrical Inspectors, for example, they can discover loose nuts/bolts on a **Visual** Inspection. Many of the pumps, for instance, that are inspected at the moment will not be Atex, therefore a Risk Assessment should be carried out by the mechanical engineer. So what **Visual** checks would a Mechanical Inspector carry out?

### Pumps in general:

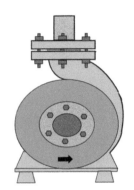

The pump will have flanges which are dealt with under a previous section (M1). Loose bolts are the usual issue. The pump will have nuts around the volute keeping it together. A **Visual** Inspection would check that all of these nuts are present and none are slack.

As shown on the left there is a directional arrow on the pump which denotes which way it **MUST** spin. This arrow should, of course, match up with the direction of the electric motor.

The Pump should have a Plant Number.

### Is the Pump Certified?

Is the pump Certified? If no Certification or, these days, Atex plate, then a Risk Assessment should be completed by the mechanical engineer. If the pump is Certified, is the **'Protection'** correct for the Zone? Is the **'Gas Group'** and **'Temperature Class'** correct for the chemical being used?

### Is the Pump Earthed?

Is the pump earthed at all? If the pump is meant to be earthed there will be an earth point on the pump body. The bolts on the volute etc should not be used as an earth.

### Has the Constant Oiler got a level of oil?

If the pump has a constant oiler or greaser bottle, is the level of oil/grease acceptable? If the constant oiler/greaser is empty then there is a leak somewhere on the unit that should be reported.

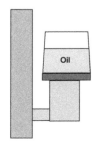

When covers are removed later for the **Detailed** Inspection, the oil seal in the pump can be checked.

# Visual Inspection
## Coupling Cover

M4

Introduction:

In the mechanical world there are only two grades of Inspections, namely **Visual** and **Detailed.** There is no **Close** Inspection, so Mechanical Inspectors do not quite come under the same Inspection criteria as the Electrical Inspectors, for example, they can discover loose nuts/bolts on a **Visual** Inspection. Many of the pumps, for instance, that are inspected at the moment will not be Atex, so a Risk Assessment should be completed by the mechanical engineer. So what **Visual** checks would a Mechanical Inspector carry out?

## Coupling Cover Metal?

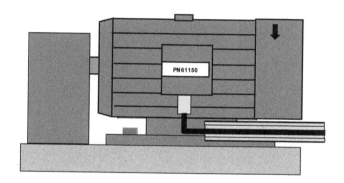

What metal is the coupling cover made of? The coupling guard should be made of non-sparking material.

Never use metals containing more than 7.5% magnesium.

## Do we Earth the Coupling Guard?

The coupling guard is sometimes earthed. The danger can be if the guard was independent and was to charge up with static electricity.

## Is the Coupling Guard Certified?

Coupling guards must be Certified for the environment that they are in. Certification of the coupling guard would be part of the pump Certification. Atex coupling guards can be obtained.

# Other titles available

## Hazardous Areas for Technicians
## ISBN 978-1-912014-95-8

Below is my other book 'Hazardous Areas for Technicians'. Both of these books are written at Technician level to assist Electrical, Instrument and Mechanical Technicians working in Hazardous Areas. The books can be obtained from Waterstones or Amazon. The book below deals with Atex and certified equipment used in Hazardous Areas and is meant to be used with the EEMUA Practitioner's Handbook (ISBN 978-0-85931-212-7). This book is intended to be used with the Standard IEC60079 which can be obtained from the internet.

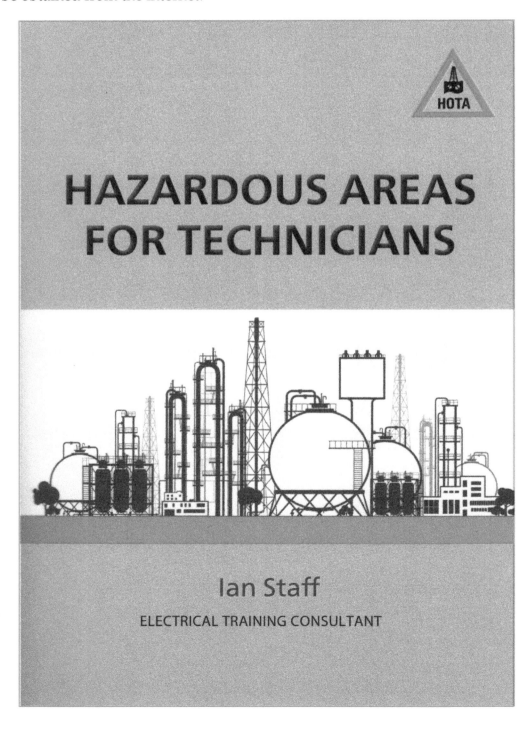

CPSIA information can be obtained
at www.ICGtesting.com
Printed in the USA
LVHW072250210720
661205LV00006B/65